HORSES & RIDING

HORSES & RIDING

Stanley Paul, London

BETTY SKELTON

Stanley Paul & Co. Ltd.
3 Fitzroy Square, London W1

An imprint of the Hutchinson Group

London Melbourne Sydney Auckland
Wellington Johannesburg Cape Town
and agencies throughout the world

First published 1974

Designed by Design Practitioners
Printed in Great Britain by
The Anchor Press and bound by
William Brendon, both of Tiptree, Essex

ISBN 0 09 118290 5

CONTENTS

TRAINING THE RIDER

SCHOOLING THE HORSE

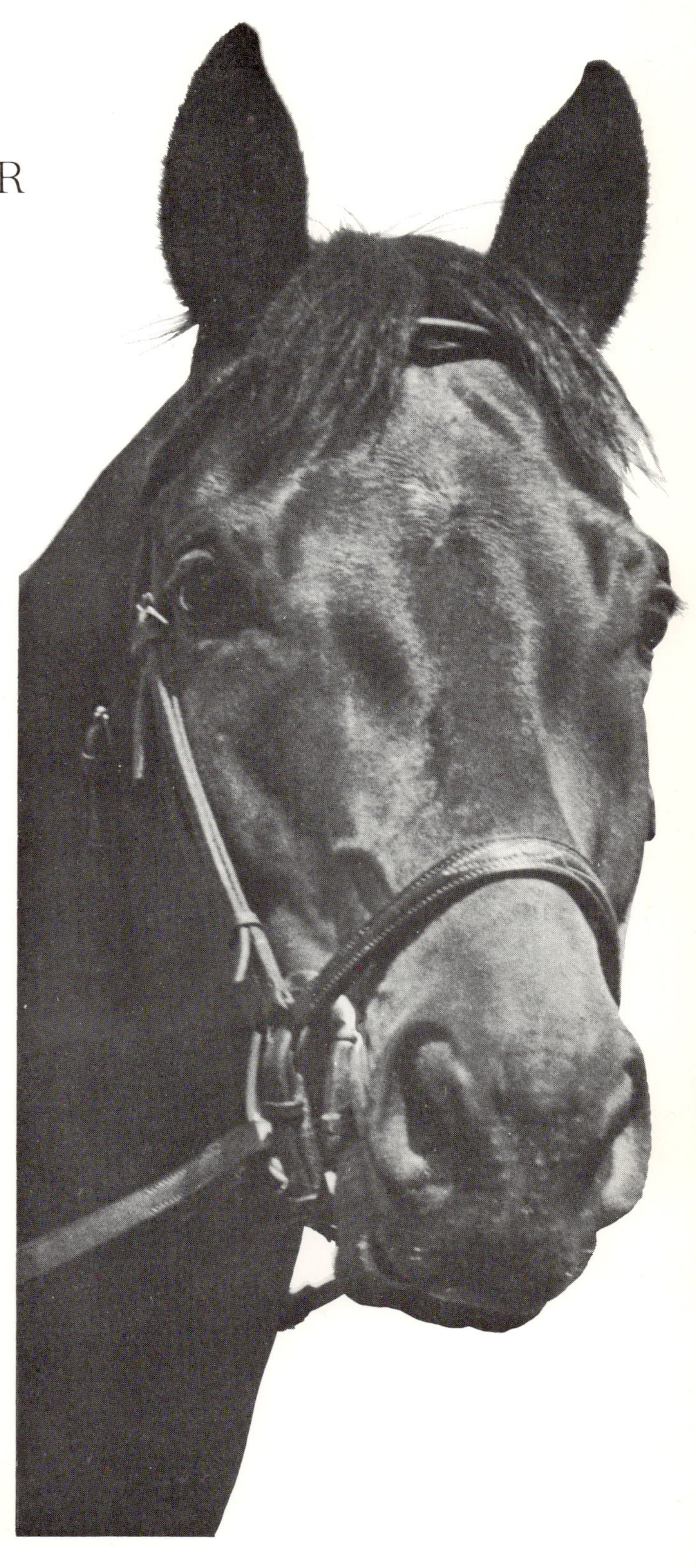

TRAINING THE RIDER

INTRODUCTION

Since the last war, riding has become increasingly popular as a hobby and many people of all ages are keen to learn more about horses and riding. Adults have taken a new interest and may decide on a trekking holiday in the mountains of Wales or Scotland or be looking for a healthy exercise in the hopes of losing an inch or two round the waist. Children, of course, always enjoy the challenge of learning to control and look after a horse or a pony. And television has given a tremendous fillip to the sport by covering international riding and show jumping events.

The less ambitious may want to be able to enjoy a sedate hack through the countryside, some may wish to hunt, and a few will go far in the equitation world. But whatever the reason for learning it is essential that the rider is taught the correct position in the saddle; only then will he be able to control both himself and the horse.

When I was young, nobody was ever taught to ride. Children were given a few hints by the groom and encouragement by the parents. There were livery stables—but none had ever heard of riding *schools*. The cult of riding had not been evolved and the art of equitation was not studied in this country. There was one exception: the Army taught soldiers to ride and produced some top class riders from the natural horsemen among its ranks. The majority of our jumping teams were composed of Army officers. But England was far behind the Continent in her approach to riding.

One of the pioneers of the present teaching system was the late Colonel Hance, who opened his own school at Malvern in the 1930's—now there are riding schools everywhere. The British Horse Society has worked to standardise methods so that there is not too much contradiction in style from one school to another—but riding instruction is a personal thing and the individual approach is bound to differ from teacher to teacher. Most good instructors attend several courses a year, picking up ideas they find useful and discarding methods with which they cannot agree. The Pony Club regularly produces excellent books on equitation, compiled by practical people, and today every child can get great help from this organisation. Adults can join the riding club movement where proper instruction and communal riding is encouraged.

Teaching methods are continually changing in an effort to make the position of the rider more comfortable for the horse. An example of this is the forward seat for jumping which has replaced the old-fashioned backward seat, in common use as recently as the 1930's.

The idea for writing this book came by mistake (like many things seem to!). I was revising, for the publishers, Captain Hayes' book *Hunting and Riding*. While sorting through the photographs (I had acted as model in many of them), I was shocked by my poor posture and position in many of the pictures—even though I was riding one of my favourite horses, Cosi. So I said, 'What we ought to produce is a book full of pictures on how and how not to ride'. Stella Robinson and Ian Gibson-Smith of Design Practitioners, with whom I was working, took up the suggestion. So it is with thanks to them, plus Cosi, and to the faith of Roddy Bloomfield of Stanley Paul, that I offer many pictures and observations which I hope will be a helpful guide for newcomers to riding, and a useful refresher for more experienced riders.

Author's Note

I have added a chapter on breeding at the request of the National Pony Society's examination committee. This is intended to help candidates taking the Society's full diploma.

FIRST LESSONS – MOUNTING

Before we can ride we must learn to get on top of the horse, and once on top we must get off again (although I have been asked, sarcastically, 'Did I ever get off a horse for meals?'!).

To start with, always try to select a horse that is not too big for the rider to mount easily, i.e. something not over 15 hands for an adult. There are plenty of useful sorts of cob which can carry 15 stone. This is where the Sections C and D of the Welsh Pony and Cob Society are so useful. Equally good is a Fell or, so rarely seen these days, a Dales pony. All these native ponies are placid by nature and suitable for teaching elementary riding. (Page 89) Nothing is more 'off-putting' to a beginner than having a fidgety horse twisting round while the poor rider hops after it on one foot trying to get into the saddle. So the instructor must choose a suitable horse for the rider. It is most important that the horse should be up to the rider's weight and that the horse has, if possible, a patient outlook on life. The same applies to a child's pony.

First the rider should be told the names of the parts of the saddle so that when he is told to take the pommel (see Fig. 2) he knows which end of the saddle the instructor is talking about. He must also be told how to guide and control the horse by the mouth and that where the bit lies there are no teeth (Fig. 1) so that jerks or hard pressure on the reins can hurt the horse's sensitive jaw. Some horses are not so sensitive as one is led to believe, as all

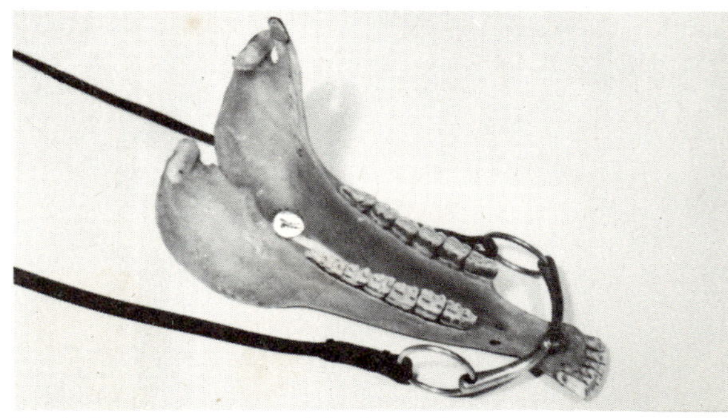

Fig 1. Lower jaw bone of the horse showing position of bit.

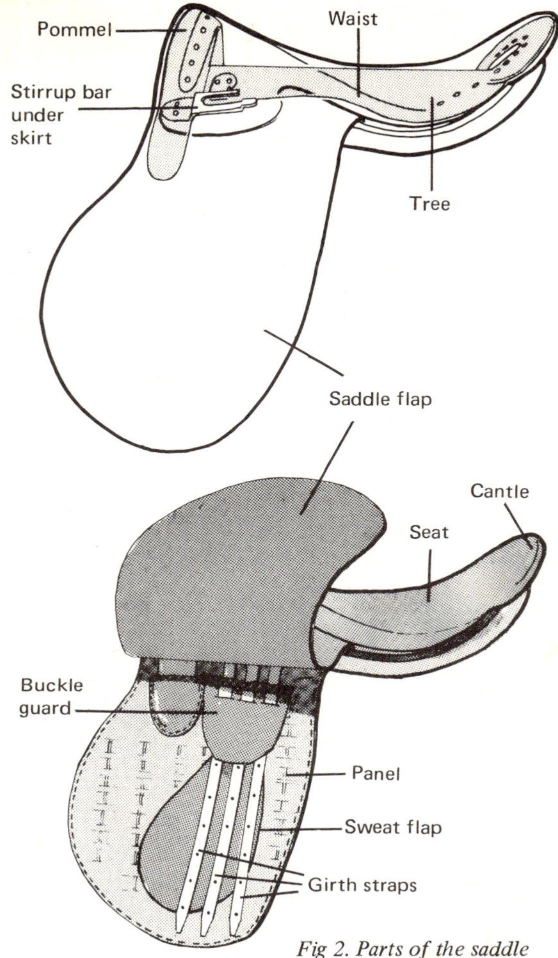

Fig 2. Parts of the saddle

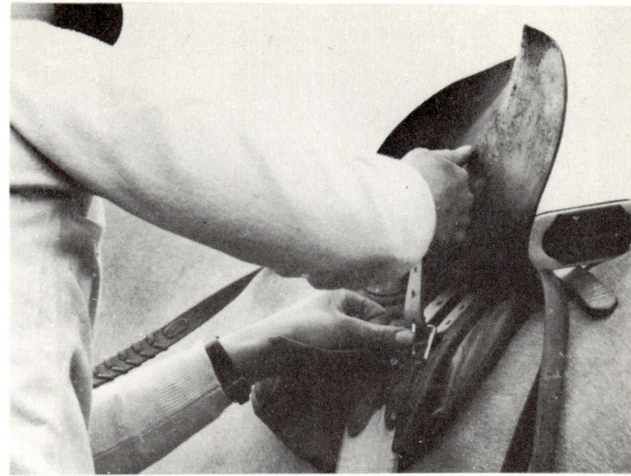
Fig 3. Tightening the girth

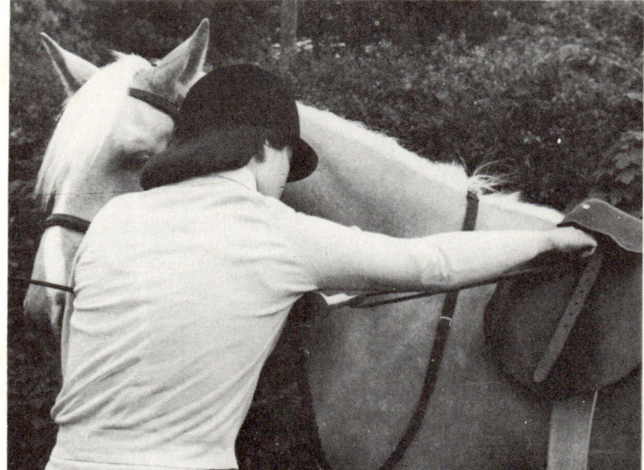
Fig 4. Measuring the length of stirrups

riders will discover. Any lack of response is usually caused by the unfeeling hands of a rider destroying the sensitive membrane of the horse's mouth. The covering of the jaw becomes insensitive and we say that the horse has a dead mouth or a hard mouth—bad mouths are always man-made!

The three aids, or signals, of riding should be explained. This can be done either standing by the horse or while the rider is sitting on the horse getting the feel of the saddle. The first aid of riding is the rider's voice, and this is the first aid a young horse is taught on the lunge. The second aid is that of the hands on the reins. These should be held gently but firmly. They should not slip through the hands and become longer, nor should the rider use the reins to keep himself steady in the saddle. The third and last aid is the use of the leg on the horse's side. A gentle squeeze says 'please go on', the lower leg down to the ankle being used in forcing the forward movement. A kick, however, is punishment for not paying attention to the first two aids of the leg or else for disobedience or naughtiness in some way.

Mounting

The British Horse Society has a very good maxim when teaching 'explanation, demonstration, execution'. First the prospective rider should be taught the drill for mounting. I know this sounds very 'army' in approach but to my mind it is important to check for safety. The rider should pass his left arm through the reins so that he can anchor the horse, and then measure the length of his stirrup leather by clenching his fist, resting his hand on the stirrup bar and pulling the stirrup iron to his armpit. (Fig. 4) This gives a rough guide to the right length for the stirrup leather. Some people have longer or shorter arms in proportion to their legs, so each person has to learn if a slightly longer or shorter stirrup leather than the accepted measurement is needed. Now the girths must be tightened until they are really firm but so that one can insert two fingers between them and the horse's side. (Fig. 3) Too tight a girth is uncomfortable for the horse, but too loose a girth is very unsafe for the rider.

Fig 5. Both reins in left hand, left foot in stirrup

Fig 6. Right hand on the pommel of the saddle

Fig 7. Swivelling the left foot round to face the head of the horse

Fig 8. Swinging the right leg over the saddle

Fig 9. Preparing to sit gently in the saddle

Fig 10. The rider in the saddle taking a rein in each hand

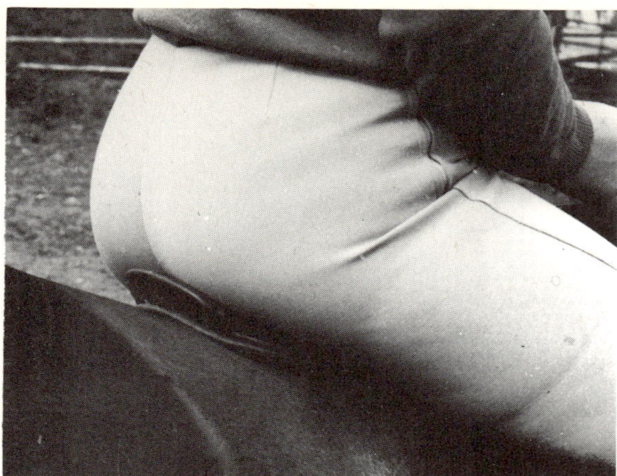

Fig 11. Saddle too small!

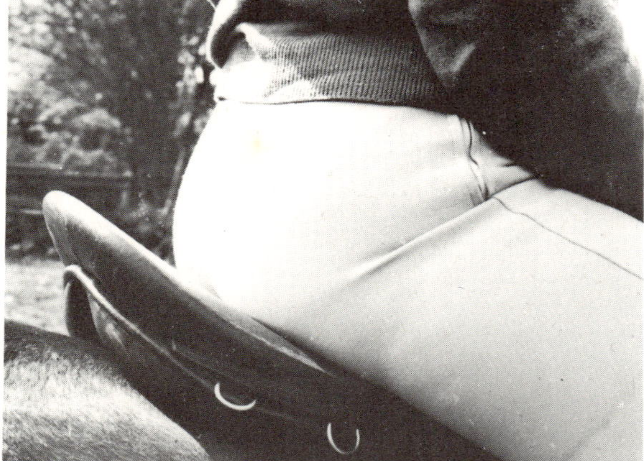

Fig 12. Saddle too large!

Fig 13. Tightening the girth when mounted– the right hand is normally used as it is stronger than the left.

Approach the horse on the left (or near) side and pat his neck. Next take the reins and a piece of the mane (or the neck strap, should the horse have a hogged mane) in the left hand (Fig 5). Face the horse's tail and take the back of the stirrup in the right hand, place the left foot into the stirrup iron and then put the right hand on the pommel of the saddle. Giving a hop and a spring, mount the horse with the aid of the right hand on the pommel. This will involve swivelling the left foot round to face the head of the horse and swinging the right leg high over the cantle of the saddle (Fig 5-10). Watch carefully while the demonstrator mounts and dismounts. When one sees a person actually performing the exercise it is so much easier to understand the sequence of the movement. Do not bump into the saddle, but lower the weight gently into the saddle, sitting with the seat bones into the lowest part of the saddle. Make sure that the saddle fits the rider as nobody can be comfortable in a saddle which is too small and sticks into parts of the rider's thighs or bottom (Figs 11 and 12). A saddle should hold the rider into it so that the rider can relax in the saddle and not be stuck on top of it. Of recent years there has been rather a revolution in the construction of saddles both from the rider's point of view and that of the horse. Hence the way of mounting has also changed somewhat in order to protect the spring tree of the modern lightweight saddle. This is deeper seated than the saddle of 20 years ago and much softer to sit in, being padded with foam rubber. We used to mount taking the waist, or cantle, of the saddle in the right hand (Fig 6) but this put undue strain on the light alloy tree of the saddle and twisted it as the rider hauled his weight into the saddle. The old-fashioned beech-wood tree, made of seasoned wood, could stand up to much rougher treatment and would take the rider's weight without going out of shape. This is why the British Horse Society wisely asked all Riding Schools to reconsider the method of mounting since it caused difficulty having to say that this saddle needs one treatment and that saddle has a rigid tree so you can mount a different way. Uniformity is always desirable if possible.

GETTING A LEG UP

Should the rider find mounting the ordinary way too difficult, the instructor can give a leg up or the rider can use a mounting block. When being given a leg up, the rider takes the reins as explained before, with the right hand on the pommel. The use of a mounting block raises the rider from the ground, so that he can put his foot into the stirrup easily. By merely putting his weight into it, he lifts his leg over the back of the saddle and mounts with much less effort. This method is excellent for those who are not so agile, but the usual method should be taught if possible, since one cannot carry a mounting block around on rides in the country!

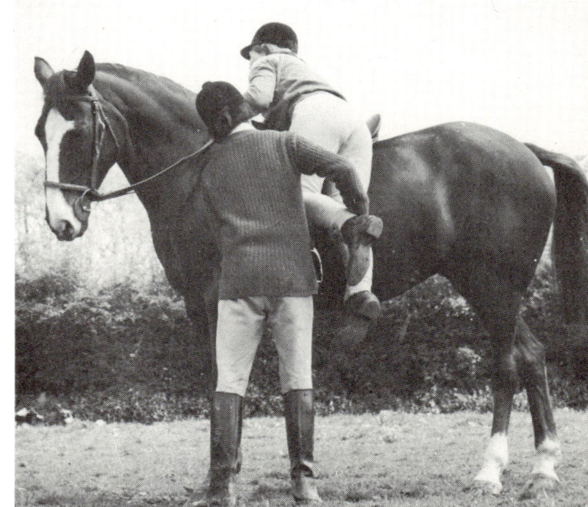

Fig 15.

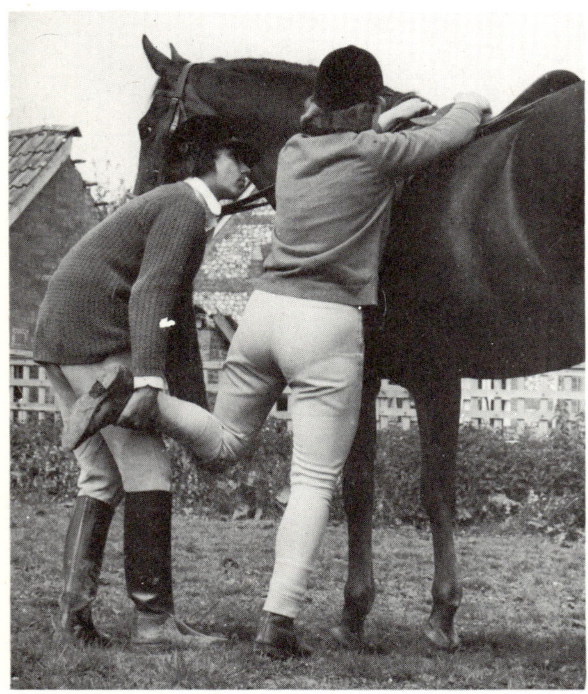

Figs 14-16. Giving a leg up

Fig 16.

HANDS

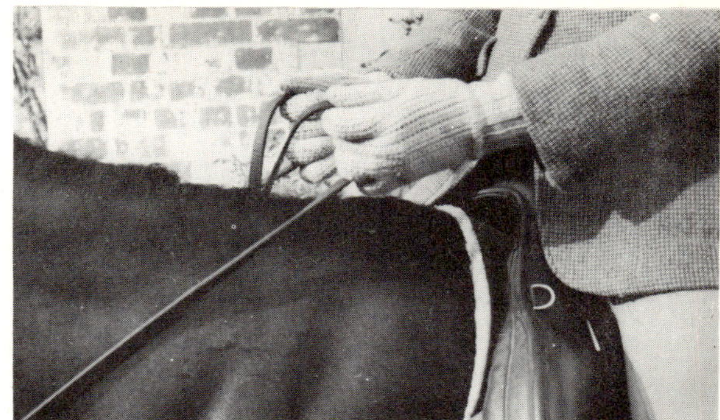
Fig 17. Holding single reins in 2 hands

Fig 18. Holding single reins in one hand

Fig 19. Constant position of the hands at a trot

Fig 20. Wrong! Reins used as 'lifelines'

Now the rider should be sitting in the saddle with the left foot in the stirrup, and the instructor will place the right foot in the other stirrup. The rider is now shown how to pick up the reins, one in each hand. They should be passing through the hands and be held between the thumb and first finger of each hand with the thumbs nearer the horse's head. That is to say, one has made a double lock for holding the reins while, with this position of the hand, the forward movement of the horse is not in any way checked. This is called allowing the forward movement, which is the give in the rider's hands when they are relaxed and giving to the movement of the horse's head—but not, of course, allowing the horse to choose his own pace. At the walk the horse's head moves slightly up and down and a good hand will move with this movement. At the trot the horse's head comes up a little and should be held in a more constant position, so that the rider's hands are more still but not pulling against the horse. This is where it should be explained that the reins are to be used to control and guide the horse and not as life-lines for the rider to hang on to or balance his body. This is where it must be appreciated that at first few riders have a perfect balance, so he should be allowed to hold either a neck strap or the saddle with his hands. I prefer the saddle since it is more stable and does not move; also, the rider is in a more natural position. The neck strap tends to move and also draws the rider too far forward if tight enough to give him any support.

I think the best way to teach the use of the hands is to liken them to the movement on the handle bars of a bicycle or the steering wheel of a car. Then the rider can follow the rein being pulled to one side and the turn of the horse's head being allowed with the other hand. All this can be done while the horse stands still.

POSITION IN THE SADDLE

Now we have to explain about the position in the saddle. The knee and thigh are close to the saddle, the leg below the knee relaxed yet perpendicular from the knee to the foot, the weight on the inside of the stirrup to drive the knee towards the saddle. This demonstrates the necessity for keeping the foot in a correct position–neither too forward, which drives the rider's weight back, nor too backward, since then it is almost impossible to keep the heel lower than the toe. There are many theories as to how one should sit, but I feel that there are certain basic principles which are universally accepted. The first is that the rider must be comfortable and be able to relax his body so that it can absorb the movement of the horse. The second is that nobody has ever taught that the toe should be kept down and the heel up (except perhaps in race riding, but the beginner will not be riding races to start with). The position of the lower leg can and should be moved according to the type of riding being undertaken, but here we are thinking only of basic riding for the beginner or for a person who rides at week ends, hacking or hunting, etc.

The rider should now be introduced to the natural walk of the horse to feel the movement which most of us enjoy so much. Then he should be told to stop the horse by ceasing the movement of his hand and putting his weight just behind his seat bones, i.e. on his soft cushion. See Fig 55. The movement of the weight of the body can then be explained to the rider by stating that his horse has been trained to answer the movement of his weight–slightly forward on the seat for walk, trot and canter, slightly back to stop

Fig 21. Leg perpendicular to the ground

Fig 22. Wrong! Weight too far forward and leg too far back

Fig 23. Wrong! Rider slumped in the saddle

and rein back, and really forward to jump and gallop. It should be explained that, when turning, the weight should be greater on the inside seat bone, but the body should stay in the middle of the horse. During all this time the instructor has been walking beside the horse, keeping a careful watch on the position of the rider's legs, feet and hands.

Fig 24. Elbows in – but sitting unevenly with one toe turned out

Fig 25. Wrong! Toes and elbows turned out

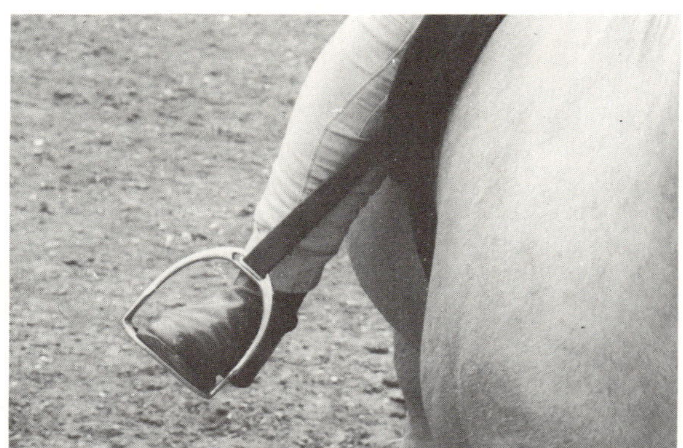

Fig 26. Weight on the inside of the foot

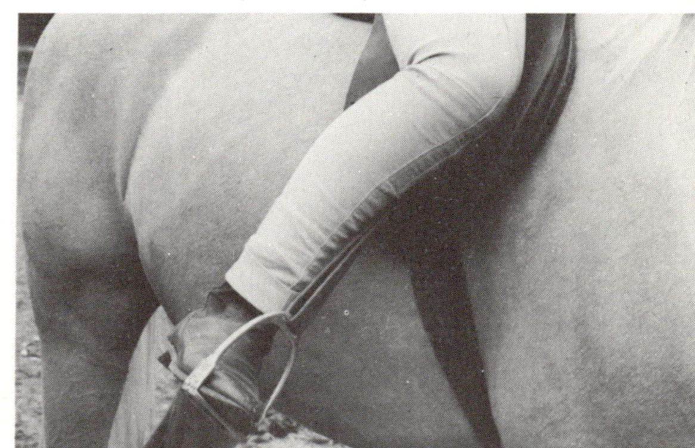

Fig 27. Wrong! Toe down

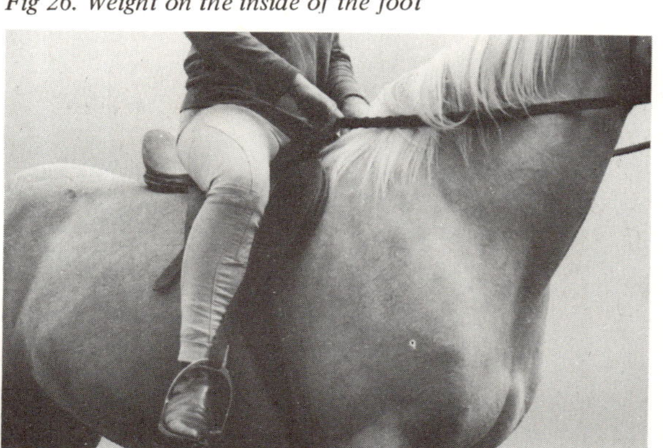

Fig 28. Wrong! Knee turned out and heel in horse's side

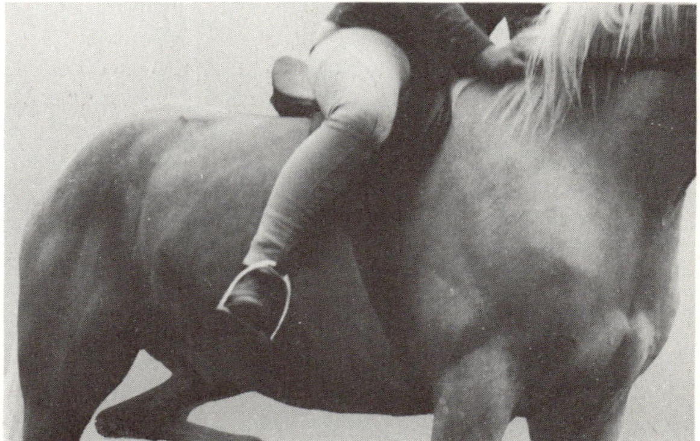

Fig 29. Wrong! Horse objecting to heel in his side

I have always been grateful to the late Phil Blackmore who, when giving a lecture to Pony Club instructors, said, 'Look at the walking position of your rider, put that on the horse, and do the best you can with it.' It is so true that a person who walks upright will naturally ride upright, while the person who stoops will also stoop in the saddle. If the instructor tries to alter the natural walking position of the rider to start with, he will produce a stiffness which makes life more difficult for both rider and instructor.

After the rider has been helped while walking, turning and stopping, he must try on his own. Most adults who want to ride of their own free will welcome the idea of 'going solo', to use a flying term, but children, or someone who has been pushed into learning to ride, may be apprehensive. This is where a horse who knows his job can help the beginner enormously and give him untold confidence. Try to train the school horse to follow the instructor then, when teaching the rider to turn right by using the right rein and the left leg, the horse follows the instructor even if the rider's aids were not very definite. This, I find, gives enormous pleasure. So we can now stop, start and turn both ways. The rider has been made to pat his horse with each hand in turn—thus practising picking up the reins and putting both in one hand.

Next, the trot can be introduced. At first, explain and demonstrate that the trot is bumpy and that there are two kinds for the rider to learn, the sitting trot and the rising trot. There is some difference of opinion here as to whether the rider should be taught to rise from the beginning or kept at the sitting trot for some lessons in order to improve his balance. Whichever way is practised, a horse who is comfortable to sit into is very important as a bumpy, uncomfortable trot will throw the rider off balance and jolt him unnecessarily. When the rise is first taught, it should be done at the walk and likened to rising on the toe and dropping on the heel when on the ground. It should be pointed out that rising too high makes unnecessary work and requires too much effort, which in itself produces stiffness from extra body effort. It is best to count for the rider 'up,

Fig 30. The sitting trot

down–up, down' or '1, 2–1, 2', the instructor looking at the front feet of the horse to get the correct timing of the forward movement of each foot. Should the instructor be riding and leading then it is very important that he does not rise at the same time, since no two horses have exactly the same rhythm in the trot, and one is apt to count as one rises oneself and not when the pupil should come out of his saddle. Never trot too far without giving the rider a rest. The way in which most people relax best is to take their feet out of their stirrups and let them hang down while walking. The clever instructor will engage the pupil in conversation as well as merely teaching him, because too much effort and lack of suppleness does not help the rider to get the beat of the rise. With talking the pupil relaxes his jaw and shoulders. I learnt this useful tip from Tony Collins when on a course at Porlock.

Fig 31. The rising trot

Fig 32. Wrong! Rising too high

Fig 33. Wrong! Leaning forward, leg too far back

DISMOUNTING

After having had your ride, you have to get off the horse. Again, there is a drill which makes this safe although there can be many variations in the way of dismounting. We will take the usual method. Put both reins in the left hand, take the feet out of the stirrups. Then, putting the weight of the body on the right hand on the saddle, thus leaving the left hand free to control the horse, swing the right leg backwards and high over the cantle of the saddle. The body will come slightly forward and the rider can swing off the horse, being careful to land with relaxed ankle and knee. He then takes his left hand off the reins and takes the left rein in his right hand, before moving to the horse's head and taking both the reins again in one hand under the horse's chin. Should you be going to take the horse back to the stable, then take the reins over the horse's head and hold both reins in the right hand close to the bit, seeing that the right rein is pulled into the horse's chin groove and that the slack of both reins is taken in the rider's left hand. With this hold the leader is in control of the horse's front end. The rider is then taught 'to run his stirrup up' as it is called; in other words, the stirrups are slid up the leathers out of the way so that they do not bang the horse's side while he is being led into his stable, nor do they catch on the doorway of the stable, which could frighten the horse and cause damage to both the horse and his saddle.

This completes the rider's first introduction to the horse. Should the pupil want to spend longer on the horse I would suggest a break–coffee or something–and then start again. Long lessons are trying from all points of view. The rider has been in a new and unusual position and both the horse and the instructor need a break to pick up their freshness again. Nothing is more tiring than continual correction. Nothing is more boring for the horse than to plod round the same old enclosed space or up and down the same old road. That is why so many riding school horses are slow in response, since it is only by dreaming that they can carry on. One often sees quite a different side of a horse when out on a day's ride or out hunting and a client will say 'I never thought Brandy could be so lively!' when he has only ridden him before in the school.

I would introduce exercises in a mild way as soon as possible to get balance and to test the control of the rider over his own body. A very favourite saying of mine is that if you can't control your own legs you have no hope of controlling the four under you. Involuntary pokes with the heel either make a horse fed up and he chucks in trying to co-operate because he gets annoyed, or they make him go too fast.

Fig 34. Both reins in one hand

Fig 35. Feet out of the stirrups

Fig 36. Body inclined forward

Fig 37. Swinging off the horse

Fig 38. Taking both reins over horse's head

Fig 39. Holding reins under chin groove

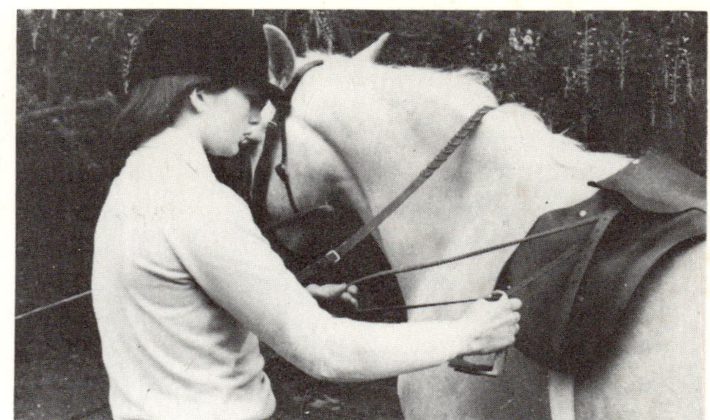

Fig 40.

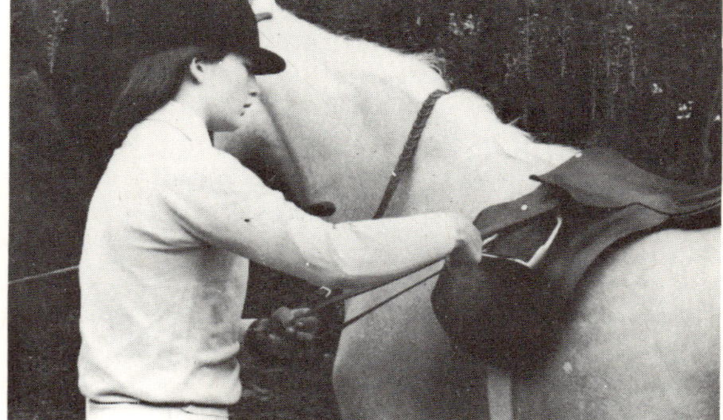

Fig 41.

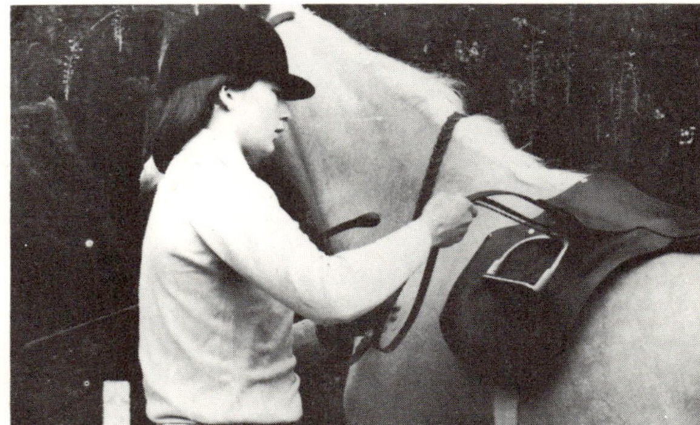

Fig 42.

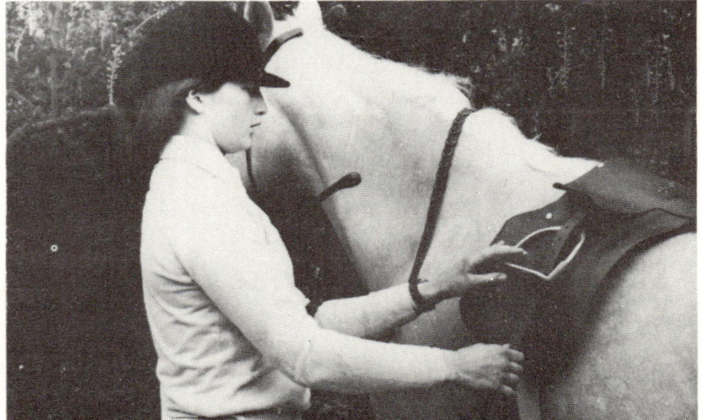

Figs 40-43. Running up the stirrups

EXERCISES

At first all exercises should be taught while the horse is standing still. Some are very simple and easy to execute. The idea behind all this work is to settle the rider more firmly into the saddle and to improve his balance and the control of his body, especially his leg position. When they can be properly executed at the walk, then introduce work at the trot. Only when the rider has a really firm seat can work at the canter be attempted.

One of the easiest of all exercises is to ask the rider, adult or child, to raise one arm to touch his horse's ears and swing round with the same arm to touch the horse near his tail, repeat changing arms. This exercise is given to give the idea of an independent seat, and to strengthen his position. This is because when a rider leans forward he tends to allow his lower leg to go backwards; likewise, when he leans backwards his lower leg tries to go forwards. When the rider feels happy doing this simple exercise, he can fold his arms and lean down towards the horse's neck and backwards, always remembering the position of the leg. (See Figs 44, 45 and 46.)

Another easy exercise is for the rider to raise his arms sideways, and swing the whole body round to look behind him. Face forward again, repeat in the

Fig 44. Rider leaning forward with correct leg position

Fig 46. Rider leaning backwards without stirrups

Fig 45. Rider leaning half way back and keeping leg position

Fig 47. Rider swinging the whole body from the waist

opposite direction. This exercise can also be done with the hands placed on the hips. This is to help develop the independent seat and supple the waist of the rider. (See Figs 47 and 48.)

Another exercise to give the rider confidence is to raise the arms level with the shoulders, keeping the head forward. The arms can be lowered and raised again, which is very good should he be inclined to sit behind his horse. Should the rider lean slightly forward, then he should cross his arms behind his back.

Here we can start exercises on the lunge. The rider holds the saddle with the outside hand and, using the inside hand, raises it above his head, then bends to touch his toe on the same side. (See Figs 50 and 51.)

Now we can turn attention to the legs. With the rider having both hands on his saddle and the instructor lungeing the horse, he raises his knees and kicks down, stretching his leg to the utmost. This draws his seat down into the saddle.

After this, we can think about the ankle: bending the toes down, swing the foot outwards and upwards and reverse the procedure, swinging the foot down and in and up. This supples the ankle.

Another exercise connected with the legs: swing one leg forwards and one leg backwards with the toes pointed towards the ground. This teaches the rider to use his legs independently and again draws the thigh well down into the saddle.

All the above exercises are aimed at suppling the rider so that his body will relax and move more freely with his horse, besides strengthening his seat.

Fig 48. Rider working on the lunge, hands on hips

Fig 49. Rider holding saddle with outside hand circling with inside hand

Fig 50. Rider with inside hand raised

Fig 51. Rider touching inside toe

Fig 52. Touching the right toe with the left hand

Fig 53. Working on the lunge, with arms folded

Fig 54. Working on the lunge–both arms raised high to straighten rider's body

An exercise to supple the waist: the rider again either working on the lunge or being led, raises his arm forward and, watching the movement of the hand, he lowers the arm and swings it round backwards above shoulder height and down level with his shoulder again. This is good for the rider who is inclined to be a little behind his horse's movement. Should the rider be a little in front of his movement, the arms should be raised from shoulder level backwards and return forwards.

Another exercise to improve the rider's seat and control of his legs is to ask him to raise his arms to shoulder height and then bend down to touch his right toe with his left hand, return to the central position and touch his left toe with his right hand, again being careful to maintain a steady position of the leg. (See Fig 52.)

When on the lunge, a very useful exercise for balance is to ask the rider to raise both his hands above his head, keeping this straight and forward, and work at the sitting trot. (See Fig 54.)

Another good exercise to develop the independent seat: ask the rider to fold his arms and work at both the sitting and the rising trot–should he tend to lean forwards, then the arms should be folded behind his back. (See Fig 53.)

THE POSITION OF THE RIDER IN MOVEMENT

At first the position of the body in the saddle should be fully explained. The rider has already worked a little at the walk and trot: now the rocking of the pelvis should be demonstrated and explained.

At the walk, trot and ordinary canter, the rider should sit in an upright position with his legs perpendicular to the ground, and a rough line should pass through his ear, shoulder, hip and heel. This is called a central position and the weight is taken on his seat bones, but for jumping and galloping the weight needs to be carried forward since the centre of gravity of the horse moves forward. When the rider wishes to halt, his weight should be placed slightly behind his seat bones, and the same applies should he wish to rein back. This enables the horse to engage his hind legs. Standing still, the rider should take the three positions which are known as rocking the pelvis.

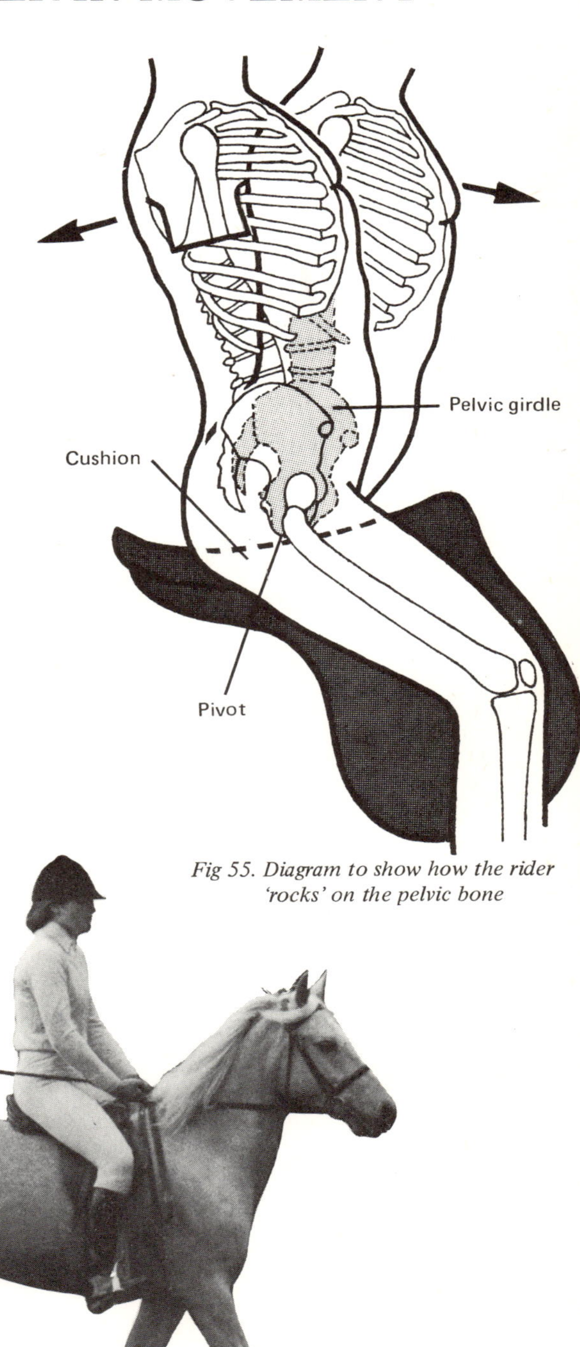

Fig 55. Diagram to show how the rider 'rocks' on the pelvic bone

Fig 57. Wrong! Body too far forward, hands fixed

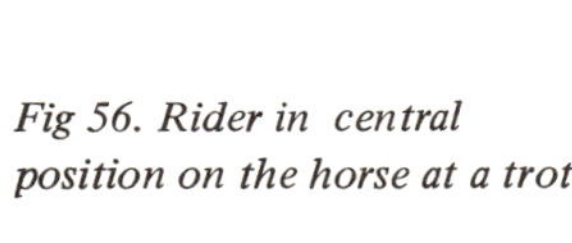

Fig 56. Rider in central position on the horse at a trot

When a beginner is first asked to trot over poles, he should take a more forward position, raising his seat bones slightly out of the saddle and forcing his weight into his knee. This is where the position of the foot is all-important in maintaining his knee in the saddle. I know that the old-fashioned way of asking the pupil to take hold of a lump of mane halfway up the pony's neck is much frowned upon, but in this way a rider stays forward with his horse. Then, should the horse either put in a short one or stand back, the rider will neither be left behind nor will he be in front of his horse. I would only suggest that this position is used in the first stage of teaching the rider the forward position over a fence. I have found that this method is most successful with small children who like jumping a willing pony: I agree it is quite disastrous with a pony who runs out or stops.

Fig 58. A more forward position to trot over poles–here the rider is pushing the horse forward with his seat and legs.

Fig 59. The rider takes hold of a piece of mane

LUNGEING FOR THE RIDER

Fig 60. Gaining the rider's confidence at the walk with one hand on the saddle to keep her balance

Before the second world war, lungeing was only used in this country for breaking horses, but after the war there was a distinct continental influence on our riding. I myself learnt continental lungeing from that great horseman and expert trainer of show horses, Count Robert Orssich. This came about quite by mistake. My daughter had won the championship at the Bath and West Show and Robert kindly came and congratulated us. I looked at him and said, 'You know I should not have won' (the pony had gone very badly) so he said: 'Let me take the pony and I will put it right'—that is how Picture Play and I went to Poplars Farm for two weeks and how I learnt continental lungeing. To my mind, this has been my greatest help, for both teaching the rider and schooling the horse, and it is now recognised by everyone to be the best way of training the horse and rider. Even the Olympic riders are lunged daily when they are in training.

Lungeing can be done to train the horse or the rider, and in this case we are dealing entirely with the rider. Therefore we must have a sensible, well-schooled horse who is obedient and calm, so that the instructor can give his whole attention to the position

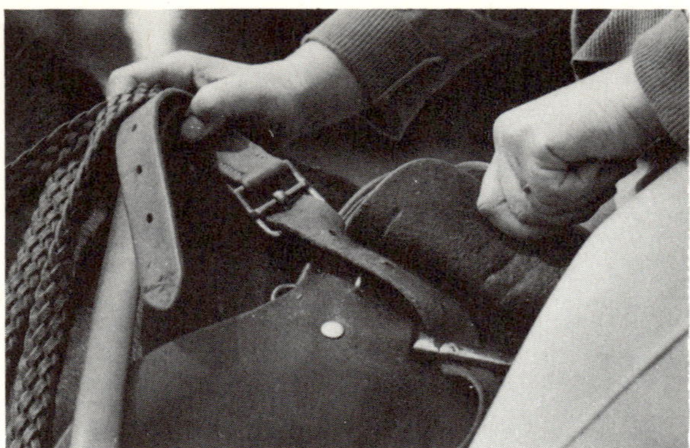
Fig 61. Pull the buckle 6in. from the bar

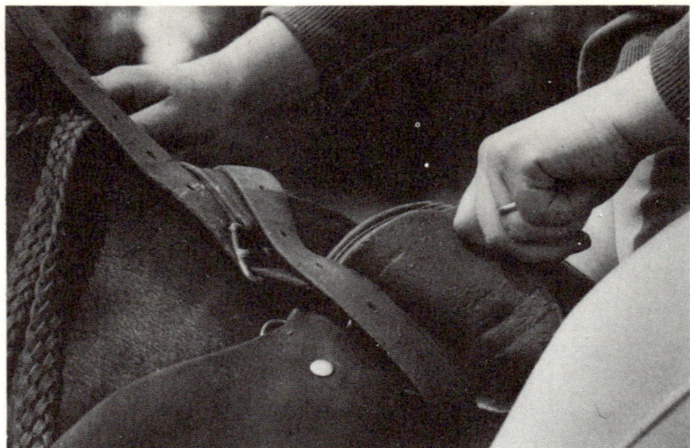
Fig 62. Cross the stirrups

of the rider in the saddle and his effectiveness in controlling the horse. Besides a quiet horse, it is essential to have a saddle which fits both the rider and the horse. The first thing is to gain the rider's confidence on the lunge. It is usual to ask him to put both hands on his saddle to help him obtain balance with the horse. It will be found that most riders ride very much better without stirrups, strange though it may seem. Should a rider feel safer with his stirrups, then I would allow him to work in this way. When enough confidence has been gained, the rider should be asked to work without stirrups at the walk and the sitting trot. Should he have stirrups on his saddle, one must remember to pull the buckle of the stirrup leather about 6in. from the bar and then cross the stirrups, placing the lower leather under the top skirt of the saddle. This is so that the leg can be in the right position without the leathers causing any discomfort to the thigh or pushing it backwards away from the saddle.

See that the rider's position is right at the outset of his work, the body being, if anything, slightly tilted forward with the weight on the seat bones and not on the cushion behind the seat. If the rider is sitting on his seat rather than his seat bones when the horse moves forward, he will be behind the movement of the horse. I like to see the sole of the rider's shoe or boot slightly as I stand in the middle of the circle, then I know that the knee is close to the saddle: in this way the rider's body should move with the movement of the horse. The object is to improve the rider's position in the saddle and to make his rhythm in unison with the horse. To start with the sitting trot is the best pace to work at on both reins. After the rider can be asked to take his stirrups and place the weight of the foot on the big toe joint, the stirrup leathers should hang in the same perpendicular position when his foot is in the iron. At first, work should be continued at the sitting trot with stirrups and then the rising trot can be introduced. When

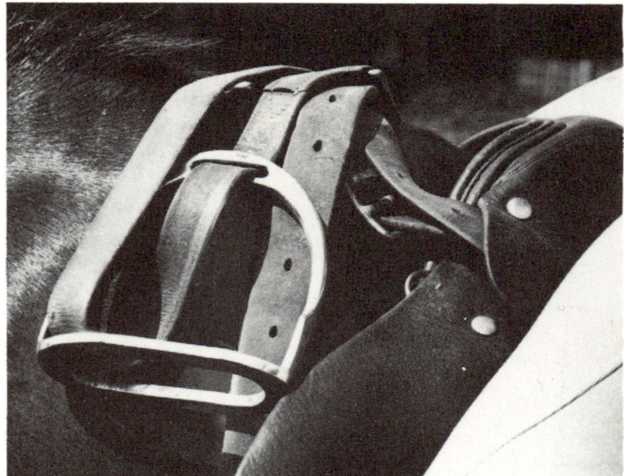
Fig 63. The lower leather under the top skirt of the saddle

better rhythm has been established, the instructor may introduce exercises with discretion and also trotting over poles, followed by small jumps.

At this stage I would suggest that 15 to 20 minutes' work on the lunge is sufficient, followed by a short hack to relax both rider and horse. As the rider improves in his aptitude on the lunge and can hold a better position, he can be encouraged to develop more control and feel in his seat and legs. The instructor starts teaching him to use his weight, voice and legs to obtain better paces and changes of pace from the horse (the rider still has no reins). The instructor must be able really to trust his horse, then he can work with less help from his whip, leaving the control to increase and decrease the pace entirely to the rider.

Work like this develops a good feeling in the rider's seat and legs, while at the same time impressing upon him that he can control his horse with a very light hand. Next, he may be allowed to take his reins and perform all his exercises almost entirely by himself.

HANDS AND REINS

When I was young it was always stated that hands were born and not made. I entirely disagree with this. First the rider must concentrate on getting a fairly independent seat so that he does not have to use his reins to keep his balance in the saddle. I feel that, if the use of the reins is explained with great care and the novice rider is allowed to ride on a loose rein to start with, then he will not rely on the rein either to pull himself up out of the saddle for the rise nor to steady any imbalance of the body. Hands should be sympathetic and yet firm–I know this sounds contradictory but it is no good a rider continually lengthening and shortening his reins. I feel that good hands come from a relaxed rider, and he should be able to maintain a steady, gentle contact with his horse's mouth equally at the rising trot as at the walk and slow trot. So many people raise their hands as they rise in the saddle because they do not have a relaxed arm. To relax the arm to the fullest extent one needs to have a relaxed shoulder: equally a relaxed elbow and wrist so that the movement of the horse can be allowed by the elasticity of the elbow and the give in the wrist–thus allowing the forward movement, especially at the walk (when the horse lowers his head at every other step). At the trot, the horse raises his head slightly, yet without altering the length of the rein the rider should be able to maintain a gentle contact. The French have a very good way of explaining the use of the hands: they say that the length of the rein from the horse's mouth to the rider's hands remains equal at all times, but the hands of the rider move to give the necessary flexion and control when riding. I would suggest that children and ladies are given small, pliable reins to hold as they find it so much easier to close their fingers on something soft and not too large, though most men prefer a larger and often a rubber-covered rein.

Exercises for the improvement of the hand can be done by asking the rider to drop the hand to his sides, raise it and drop it in front of his saddle, making him realise that the weight of a relaxed wrist can open and shut his elbow–therefore there is no need to raise the hand when rising. The movement of the wrist can be likened to the hinge of a door opening and shutting when asking the horse to halt or reduce pace.

Later on in the training of the rider it may be necessary to introduce a double bridle, i.e. two bits in the horse's mouth. The bridoon or snaffle raises the horse's head while the curb rein flexes the horse's head from the poll. Over use of this rein can cause the horse both to drop his head and to carry it behind the perpendicular. At first the rider should get accustomed to holding four reins in his hands with the curb rein very loose. It is a great mistake to use a double bridle with both the curb rein and the bridoon rein at the same tension.

At first I think a rider should be shown that there are many ways of holding the reins in his hands so that he can try out the effectiveness of the application of the bits in the horse's mouth and understand what he is asking in the positioning of the horse's head. First, if he uses the curb reins in one hand and the bridoon reins in the other, he can operate which bit he wants. He should try to elevate the horse's head with the bridoon rein and leg as he moves forward and then to flex the head from the poll, feeling and realising as the horse drops his head and gives to the curb pressure. Then he can try holding the reins in two hands, the curb rein running through one hand up and over the palm with the bridoon rein running down and under the palm, closing the fingers on the reins. Either rein can then be used by turning the wrist. But to my mind this causes a stiffness in the hand. This used to be called the French way. In the English way, the curb rein comes round the little finger with the bridoon below the third finger or, if you prefer, between the second and third fingers (the

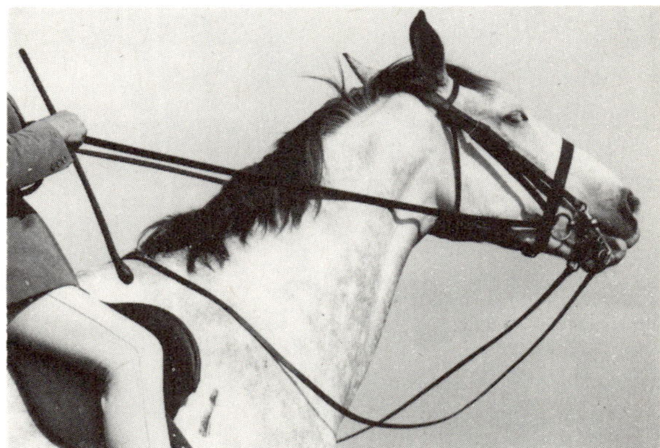

Fig 64. The bridoon rein raises the head

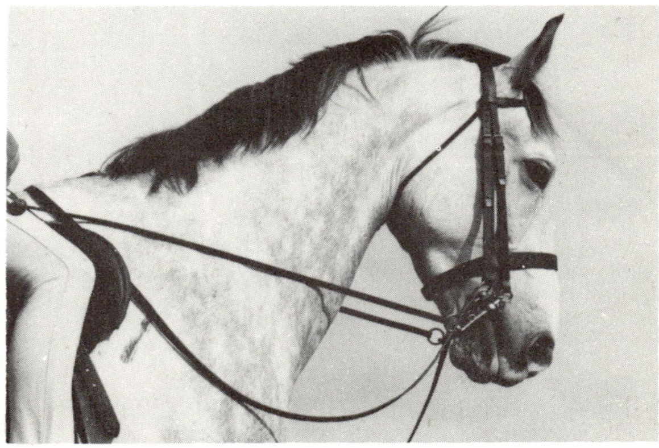

Fig 65. The curb rein flexes the head

2-finger space between the reins does help the less experienced rider to use the reins independently). Thus with the use of the hand the greater pressure comes on the curb rein. Now we come to the Continental method of holding double reins, which is the one usually accepted these days. The snaffle rein comes on the outside and the curb on the inside of the hand. Thus the greater influence is on the bridoon rein which suits most horses since they tend to have too low rather than too high a carriage. But it must be realized that both are correct and either can be used according to the needs of the particular horse's head position. The English way tends to lower the head while the Continental method raises it. I can never understand the controversy over the desirability of one or other method when surely only the position of the horse's head can determine which should be used.

There are horses with conformation and rightly set-on heads to which flexion of the head from the poll comes naturally. Others with thick jawbones find it both difficult and, even, painful to flex. The usual cause for this is too small a space between the jawbone and the neck which leads to a pinching of the perotic gland. This causes discomfort to the horse and he will, in all probability, start to throw his head up and down to relieve his aching neck. Such horses will cause the rider a lot of trouble no matter how gentle and good his hands may be.

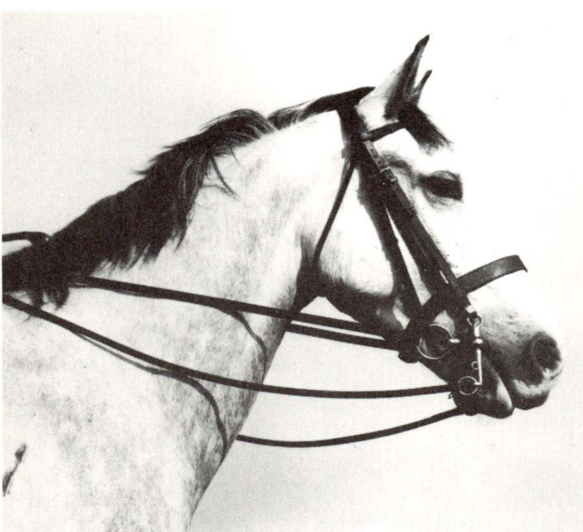

Fig 66. Curb rein very loose

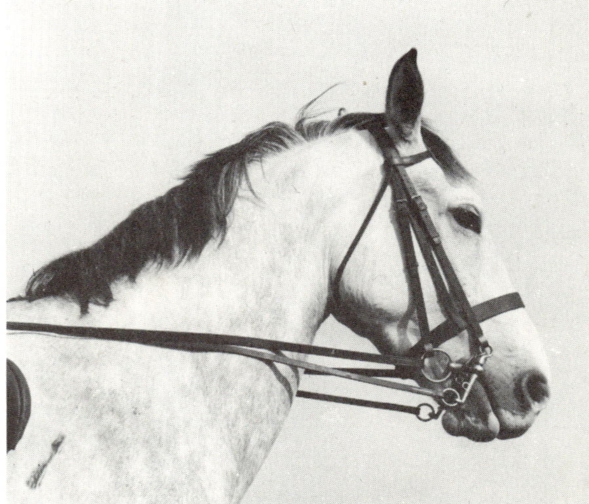

Fig 67. Wrong! Curb and bridoon at the same tension

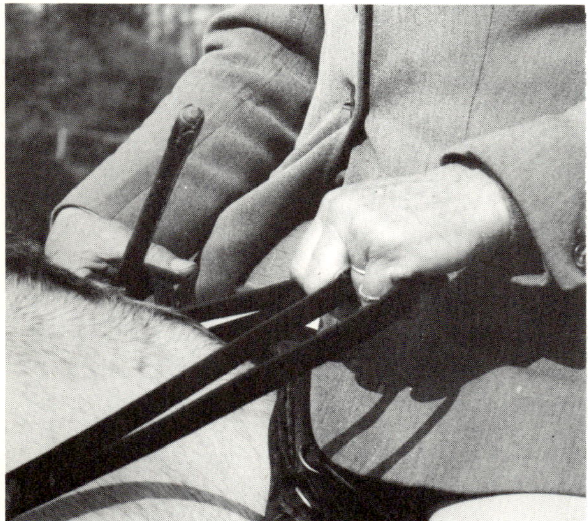

Fig 68. The English method–curb rein on the outside

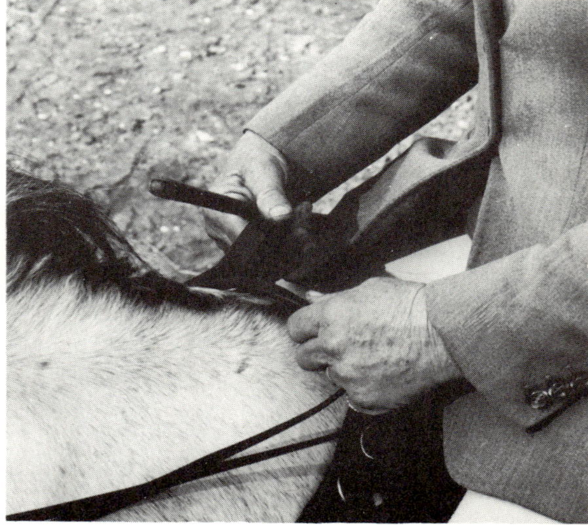

Fig 69. The French method–bridoon rein on the outside

PROGRESSIVE WORK

Fig 70. A secure seat

Once the keen rider has developed a fairly secure seat and enjoys working at improving his riding rather than going for hacks, more work can be done on the lunge. Riding well-schooled horses, he must learn how to reduce and increase pace smoothly—in other words, to make good transitions up and down from the walk to the trot, the trot to the canter, and also to learn something about the paces of a horse.

There are three paces of walk, three trots and three canters. In other words there is the ordinary walk, the extended walk and the slow walk, which can also be called the collected walk. The fully extended and collected paces are difficult to obtain except by an advanced rider on a very well-schooled horse. Therefore it is better to work at the more ordinary paces (the ordinary walk, trot or canter) and what are known as the strong paces, because real extension and real collection is very advanced work.

The other type of work to be considered is riding on straight lines and circles. One of the easiest ways to ride on straight lines is to practise riding from a given point to another marked point across the field, telling the rider to keep his horse's head straight and trying to see, say, a tree, a flag or any definite object directly in the middle between his horse's two ears. The instructor should then stand behind him, correcting the position of the horse's quarters, urging the rider to use his legs to control them. When the rider turns, then the instructor can watch his pupil from the front, keeping a good look out for the horse's legs to track up correctly (that the footprints of the near hind should fall in a straight line and slightly overtrack the footfall of the near fore, the same with the off hind and off fore).

When working on a circle it is essential that the rider makes his horse describe a circle with his whole body: the flexion of the head and neck is controlled by the hand and the bend of the quarters is guided and controlled by the rider's outside leg. Working on an average-sized circle, the rider can ask the horse to increase and decrease its size, still keeping correct flexion and bends. This is done by placing the horse on the circle and asking him to increase its size by the use of the inside leg over the girth pushing the horse to the outside and indirect rein, which can be used slightly outwards. The direct aids of the inside hand the outside leg behind the girth remain static. After this exercise, describe one or two circles on the larger diameter and then reduce the horse's circle back to the original one. Repeat on both circles. If you wish this work to be very exact, I would suggest you mark out two circles of, say, 10 metres and 15

Fig 71. When working on a circle the rider makes her horse describe a circle with her whole body

Fig 72. Note the rider's head is not quite straight, obviously she is talking to the trainer

metres diameter, which can easily be done by one person holding a long lungeing rein in the centre while another person marks the circle with sawdust and repeats the exercise on the larger circle as well. Thus when the circle is correctly ridden the horse will be moving the sawdust with his feet the whole time. This movement leads us to the beginning of two-track movement, in other words, when a horse moves sideways and forwards. (See page 35.)

Not everybody wants to work at more complicated movements; but practically everyone wants to be able to jump.

Fig 73. Flexion of the head and neck is controlled by the hand

Fig 74. The bend of the quarters is guided by the rider's outside leg

JUMPING

Fig 75. Taking a lump of mane and keeping the elbows bent–Note the neck strap and the position of the rider's hands.

Jumping is not a necessity for every rider; many people can enjoy hacking without the horse ever leaving the ground. Should the novice rider show any signs of nervousness, then I would leave the question of jumping until he has some balance and more confidence in both his instructor and his horse. On the other hand, should child or adult show no signs of apprehension, and have a relaxed position in the saddle, and wish to jump asking 'How soon can I jump?' or giving a positive reply to the question 'Would you like to jump?', then I can see no reason why the jumping position and trotting over poles should not be taught at the first lesson, since the rider does not need to rise correctly at the trot to be able to jump. If all goes well with the position trotting over poles, then I see no harm in giving the rider a final hop over a 1 ft. high pole. I know a lot of people will disagree with me entirely, but I feel that it introduces jumping as a part of riding and I know it gives children a tremendous thrill to be able to say, 'Mummy, I jumped!' To the adult it gives a sense of achievement and the feeling that really it is not as difficult as it appeared to be.

I would make the introduction to the jump very easy by explaining the rocking of the pelvis in the three positions, stressing the forward position to jump, in the centre or upright for the walk, trot and canter and backwards on the seat rather than on the seat bones to halt and rein back. This is when the weight of the seat bones is just raised out of the saddle. The rider can maintain this position easily if he takes a lump of mane in both hands, half way up the horse's neck, and keeping the elbows bent, he will hold the correct forward position. The instructor leads the horse and walks a little until the position is understood. The rider then sits upright and rests for a few moments, then walks over the poles two or three times followed by trotting over them. Again the rider

Fig 76. Jumping a ditch – the rider's weight is taken on the thighs and stirrup leathers

sits up and rests after each trot over the poles. Should the walk and trotting over the poles go well then the rider may trot over a higher pole.

I would suggest that three poles are used, two on the ground and one raised a little off the ground. You might call it 'jumping without tears' but at least the rider feels the movement of the horse and the sensation of the lift from the ground. Those who are against this method can argue that the rider has no control. This is agreed but in any case my rider will either take his mane or be lunged without reins for some time while jumping so that my horse will not have his mouth hurt if the rider loses his balance and uses his reins as lifelines. One can argue that a neck strap would be better for this purpose, but unless this is tight and halfway up the horse's neck the rider would not attain such a good forward position; nor is a neck strap so steady to hold on to as the mane.

When teaching jumping you must have horses that are well-schooled and will willingly go forward over cavalletti. In my opinion, when working with a novice rider who wishes to jump, one works first at the rider's position in the saddle and then on strengthening his seat. Both of these are best done on the lunge. Jumping lessons should be private lessons if possible as the instructor has his work cut out in correcting the position of the body, legs and hands. Then progress to working with a single bridged rein holding a forward position at the trot, changing direction, reducing pace to a walk, increasing pace to a trot and finally working at all three paces, walk, trot and canter, with frequent rests for the rider.

When resting, the rider should take his feet out of the stirrups, pointing his toes downwards to relieve the thigh muscles, and swing his legs backwards and forwards. By sitting upright the rider's back is rested; by the toe being held downwards the thigh muscles are relaxed. While the rider has his feet out of the stirrups he can work at relaxing his ankles by circling his feet both ways, the toes being dropped down. First circle up and outwards, then up and inwards towards the horse. It is important that the rider maintains a relaxed ankle when jumping so that the action of the ankle works as a shock absorber when landing over a fence.

The more independent and strong the rider's seat is, the easier it is for him to maintain a steady and 'giving' hand. 'Giving' so that he can allow the horse

Fig 77. Peter Robeson riding First Time *at Windsor*

to lower his head on approaching the fence in order that he can look at the ground line of the fence (fences without ground lines are always more difficult to jump). Then the horse raises his head to get over the jump and lowers it again on landing. The rider's weight is taken in his thighs and on his stirrup leathers. The ideal is to have the stirrup leather perpendicular to the ground but a lot of riders do swing their legs backwards when jumping. Of all the show jumpers, Peter Robeson is a great stylist: a beautifully relaxed, calm rider: a perfectionist. He has produced such horses as his famous *Craven A* and *Firecrest* and now *Grebe*. There are many helpful photographs of riders going over fences at the back of *Riding Logic* by Meuseler. They were taken many years ago but the positions are the same today. The problems, likewise, are the same. We have come to accept and appreciate the Italian forward seat here in England but there is still a tendency for many riders to get left behind the horse's movement when jumping, especially over spreads.

When working at the improvement of the rider's seat, work without stirrups should not be neglected. Should a rider lose a stirrup and not be able to retrieve it before the next fence, he should be confident in his own ability to jump without stirrups and be able to ride his horse on into the fences so that his horse does not realise his rider is in trouble, thus destroying the horse's confidence in his rider.

Many years ago a pupil of mine was jumping at the International Horse Show, then held at the White City in London. She lost her stirrup at the first fence and jumped a clear round with one stirrup. I had always made the children I taught jump first without one stirrup then without the other—you never know which stirrup you may lose—and finally without either stirrup. This year, on television, they showed bare-back jumping at the International. It was a good example of the need for balance and the ability to ride forwards with an independent seat, besides showing a lighter side of show jumping.

At first, when jumping in the ring, the ideal is for a rider to have a good school master—an experienced horse to help him learn the job. Then the rider must try to become an active horseman and be able to help his horse. This needs experience and an understanding of stride and distances. The rider and his horse need to practise different easy combinations of fences in order to get the feel of the number of strides his horse will take between fences. The fences should be low so that the horse and rider have complete confidence and they both can concentrate on the stride of the horse. Will this particular horse need to jump in on a long stride to get his jump out in the right position or does he need to pop in and stride out? Will he take three strides where another may only take two, etc? It is a case of knowing your horse by trial and error. How often if standing at the collecting ring one hears even the great riders come out saying, 'It was my fault, not the horse's.' One can learn so much from the great riders, listening to what they say and watching them. If one goes to a horse show the riders may be seen striding the distance between the fences. Each horse's own length of stride differs a little and the rider must get to know how he needs to be ridden at the fences. The good horseman is one who can alter his riding to suit his horse's capabilities: to help when necessary and to sit still and instil the will to get over the fences. The greatest thing of all is not to get left behind the movement of the horse and so put the weight down on the horse's back when in mid-air, thus making him drop his hind legs, hurting himself and shaking his confidence both in himself and in his rider.

So the rider must be keen enough on jumping to work hard at his own riding, always expecting to suffer aches and pains as a result. And the horse must like jumping. So when choosing a horse for this type of riding, the horse must look as if he loves jumping: he should approach his fences with his ears cocked forward and never jump for fear of what pain might be inflicted on him if he didn't.

MORE ADVANCED WORK

Fig 78. The joy of riding and schooling horses

The rider having developed a more independent seat, he can be asked for better use of his legs and reins. Not just increasing and decreasing pace and change of direction, but the positioning and flexion of the horse's head and body which is required in carrying out certain movements when showing or doing dressage. This is where the hand and leg must be in complete harmony.

The use of the leg is all-important since it creates the impulsion or driving power of the horse. The rider's seat is firm enough in the saddle, so that he is able to keep his legs close to the horse's body, so close that he can feel every movement or swing of the horse's rib cage, yet not give any aid with his legs unless he means to do so, and that aid can be just when, where and how strong the rider needs. The rider's hands must develop a more sensitive, kind and giving aid, almost a loving touch which can give and yet not lose contact with the mouth: by this I mean that as soon as the horse answers the hand, the hand gives and relaxes, rewarding the horse for answering to the aid of the rein. Should the horse start to lean on his bridle, he must be taken off his bridle, by gently moving his bit in his mouth, or by raising one hand slightly so that the bit or bits slant in his mouth. This is where the rider's seat, back and weight want to be used to help his hand and legs, so he can use himself to the utmost, balancing his horse really between his hands and legs. Here a rider does need help from an instructor who knows about training horses as well as equitation, because this kind of riding is positioning the horse correctly, be it on straight lines or curves, correcting him if he is wrong, or should the rider ask too strongly or over correct a horse then the animal must be given the counter aids to get him right again.

Having established this feeling of the rider for the movement of the horse, the more ambitious person can go ahead and try to ask for more exact movements from his horse. This necessitates some kind of area in which to work. It need not be an expensive covered school, though in the English climate they are lovely to work in either to keep dry and to be able to work at night or in frosty weather. It can be just an area of about 70ft by 140ft or any convenient rectangularly shaped piece of ground. Markers should be placed to enable the rider to judge his distances and to provide a point from which he leaves the outside of the area for such movements as changing the rein on the diagonal (it is amazing how soon one wears a track round the outside of the area or across the centre on any grass surface).

The rider will now have some plan in mind for his own and his horse's future. It may be dressage only, or it could be schooling to get good dressage marks for eventing, or it could be just the joy of schooling horses and riding with no particular plan in one's head.

Artificial Aids

I have not discussed artificial aids at all since, in my opinion, they do not concern the rider until he is doing more advanced work and, as a fairly good rider, has developed judgement in his dealings with horses. Now he should consider two important artificial aids to help his legs (mine are short and weak!) unless he has good long, strong ones in which case there will be no need for him to use the whip or spur. The whip should be employed to enhance the lower leg aid and should only tap the horse just where the rider's heel should be used, i.e. just behind the girth. It can be used very lightly and is often very useful to wake up a lazy horse or draw an inattentive horse's mind to his work. A little tap will say to him, 'Come on, stop thinking about other things and listen to me!' There are many different whips on the market, some heavy, some light; some long, some short. I think each has its own uses. Heavy sticks I never like as they interfere with the movement of the rider's hands on his reins and are cumbersome. A light whip is better for the opposite reasons, namely that it takes less to hold and does not make a difference between the hand with the whip and the one without. A lot of trainers carry two whips, one in each hand. This is a good idea as, used on top of the quarters, they do activate the hind legs although the touch of the whip, if long enough, will be found equally effective if used low just about the second thigh. But the rider is inclined to lose the horse's mouth in this case whereas if used as in the Spanish School, on top of the quarters, the rider is able to keep a better contact with the horse's mouth, as in the *piaffe*. Here I owe what I have learnt about this pace, and the use of the stick, to Vituosa, a Lippizaner belonging to Fred Keil. It was Fred who first made me interested in these beautiful horses and gave me countless books on them and their training. When I visited Vienna I felt I knew the famous Riding School and recognised so much I had read and the pictures I had seen, that I felt at home at once.

Fig 79. Fred Keil on Virtuoso

Short sticks are merely the correct dress for children riding in the show ring. They are ineffective as children cannot reach the hind quarters without too much body movement. The best sticks, I feel, are small hazel switches taken from the hedges. They can be cut to whatever length or thickness you personally like or require and are very cheap and easily replaced.

The spur should, in my opinion, only be worn as part of formal riding dress, when they can have their rowels removed. There are some horses who pay more attention to a blunt spur than just the heel but it should only be used when a horse does not obey the leg aid promptly and accurately, in which case the extra aid should be applied just behind the girth. When used too far back, some horses will kick against it. The correct use of this artificial aid will produce a finesse that constitutes elegant riding and brings instant obedience from the horse. Ultimately a well-schooled horse should need neither whip nor spur. Some people will disagree with this statement but no horse is perfect and, like children, they dislike doing certain things and so think up ways of getting out of doing exactly what they are asked to do. Or the horse may not have had a good training before coming into the possession of this particular rider, in which case artificial aids such as the martingale can be used to help keep a horse's head down. But this aid or contraption should only be used as a means to an end to get the horse out of a bad habit. No martingale is allowed in dressage nor in the show ring except for working hunter or pony classes where jumping is included, or in straightforward jumping classes. In fact I feel sure that a judge takes into consideration the use of a martingale when awarding marks for style. I know I do as I would prefer a horse which did not rush his fences with his head up. This is where all judges' personal experiences play a part and one remembers, 'Oh dear! That looks like so and so with whom I had so much trouble. No, I wouldn't like to take that home. I feel that horse there . . . ' Yet I agree that anyone can have great fun and come to love a difficult, impetuous horse. They are usually so game and generous.

Martingales

There are three main kinds of martingale. A *standing martingale* is used to control the horse's head by not allowing him to raise it beyond a certain angle. This is determined by the length of the strap from noseband to the girth. The *running martingale* is put on the reins to keep the angle of the rein down on the bars of the horse's mouth. In the old days, it was said 'A standing martingale for the horse's head, a running martingale for the rider's hands', which I feel still applies today. Yet a running martingale can also be used to keep the reins from getting over a horse's ears, having the same use as an *Irish martingale*, for a horse who throws his head around playfully. There is also the *bib martingale*, much favoured in racing stables.

Working on much the same idea is the *Cheshire Martingale*, which is often used for schooling by the jumping fraternity. This, however, is attached directly to the bit rings.

The Running Reins

These are long reins running from the horse's girth through his bridle and back to the rider's hands. They are useful for asking a horse to drop his nose and flex from the poll and should be used rather like a double rein in the rider's hands, holding the ordinary snaffle rein more firmly.They can be very severe reins if not used with a sympathetic and giving hand. As soon as the horse drops his nose and gives, then the pressure should be relaxed: it is easy to overflex a horse and get his head behind the perpendicular.

Some books include the most strange artificial aids: such things as bandages, which are not allowed in dressage and which are leg protection to my mind. They are necessary on young horses which are unbalanced in their movements and therefore likely to hit or strike their legs, rather like a child tripping over his own feet. Also cruppers, which I can never see as anything but a bit of saddlery for keeping a pony's saddle in place when the animal has a badly-shaped wither. They are very necessary in the case of a donkey without withers behind which to place the saddle. Cruppers can also be used for animals working on steep hills such as when deer stalking in the Highlands of Scotland, for timber hauling or when fencing poles are being taken up or down hill on panniers.

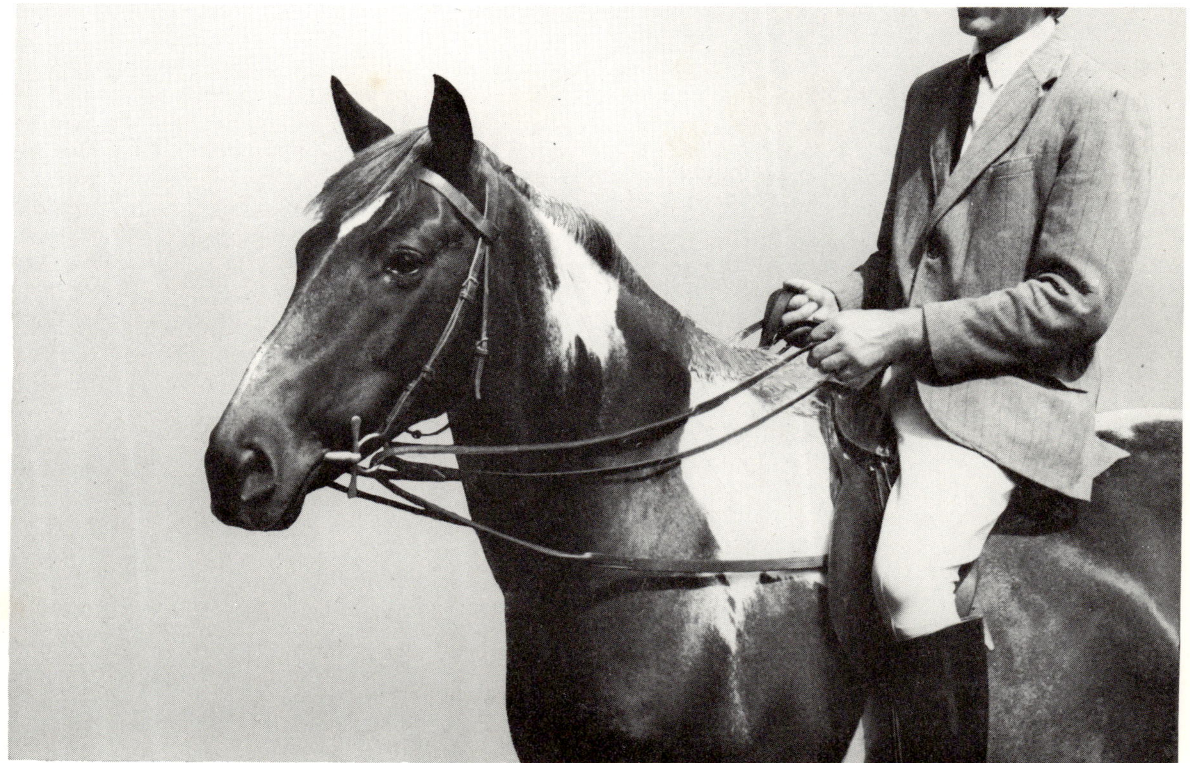

Fig 80. Running reins fixed to the girth

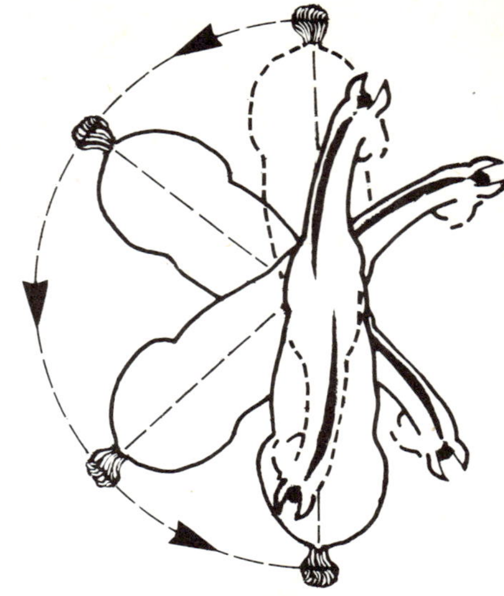

Fig 81. Pivot on the forehand

All new movements, and some schooling, should be continued in a snaffle bridle as too much use of the double bridle will make the horse less ready to be bright and responsive. Additionally, in the hands of the less experienced rider there is usually a tendency to ask for too much flexion or collection with no loss of the free forward movement and thus the horse comes behind his bit.

Having learnt how to use a double bridle, the rider should employ it on occasions to keep himself and his horse in practice while the snaffle is used for nearly all work and schooling. The more the rider can get his horse to use his back and hind legs, the lighter the horse will become in front, i.e. in hand and on the reins. There is nothing that gives greater pleasure than riding a well-schooled horse who enjoys his work and responds to the slightest touch of hand or leg, and therefore when out hacking or in the schooling area one will continue to ask the horse for an exact movement, a quick but calm response to his rider's aids.

Once the simple changes of pace from walk to trot, trot to canter, up and down, have been mastered, the rider can start to ask for some bending of the body of the horse. So far the rider has been using the diagonal aids of opposite hand and leg. Now he can learn the full use of the lateral aids, i.e. the same hand and leg, which to my mind are the forcing or corrective aids. The horse is shying, his head turns towards the object and his quarters swing away. The rider corrects the horse's position by putting the head straight and the quarters into line with the head, using the same leg and hand. These are the same aids as are used for the *pivot on the forehand*, a movement which is not much favoured by the dressage fraternity at the moment though it used to be used a great deal in Pony Club tests. I feel it can still be useful, mostly to enable the rider to gain control of the horse's quarters, since they are moved round by the rider's leg behind the girth. The rider's active leg pushes the quarters while the outside leg over the girth keeps the impulsion and stops the horse from stepping backwards. The inside hand guides the horse's head round to follow the quarters while the outside hand resists too much forward movement since the horse should take only one step forward during the whole movement. It is best to practise the pivot coming into a halt from a walk with the horse on the bit so there is still impulsion in his body, then do a half turn into the opposite direction and walk on. When control is very good, try a complete pivot and continue to walk on in the same direction.

To gain control of the forehand, *a pivot on the hocks* can be practised. This is the use of the diagonal aids, i.e. opposite hand and leg, and the easiest way to learn is to ask for a half turn into the centre of an area and halt from the walk. This is the beginning of the very advanced movement of the pivot at the canter. To gain more control of the horse's body the rider should practise such movements as the shoulder in and out, *renvers* and *travers*. These four are very closely connected, only the bends asked for being

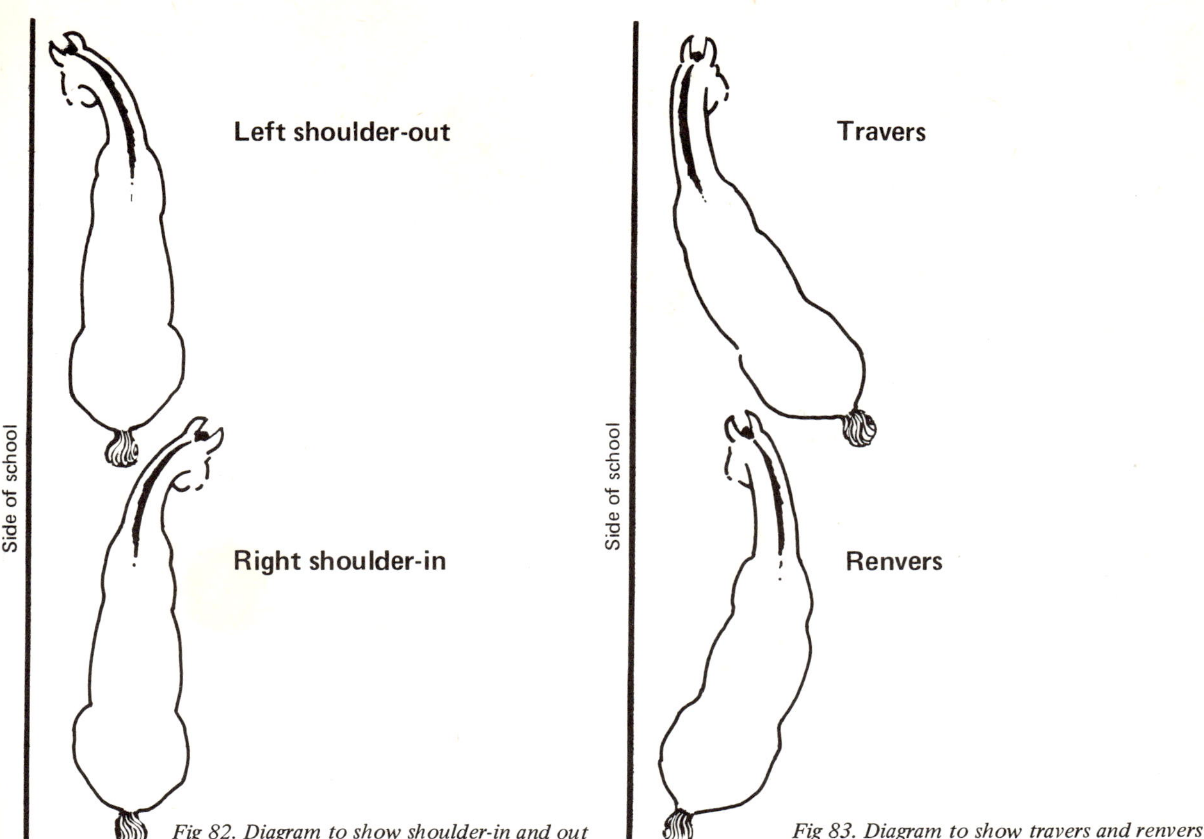

Fig 82. Diagram to show shoulder-in and out

Fig 83. Diagram to show travers and renvers

slightly different. The shoulder in or out means that the rider asks the horse to bend round his inside leg. Keeping the horse's quarters in a straight line, the forehand is brought off the track so that the head and neck are bent inwards towards the centre of the school. When tracking right, the off fore makes a path of its own while the path of the off hind falls into the path of the near fore, leaving the near hind to make yet a third path. That is to say, the head is bent against the movement of the horse. For shoulder out, the horse's head is bent towards the wall or fence on which the rider is working. It is contended that at first the horse's legs are in three paths and later in four paths when the horse is crossing both in front and behind. This, to my mind, is the half pass with an incorrect bend.

I see that *travers* and *renvers* are asked for in the new tests arranged by the Association of British Riding Schools, so I feel they should be explained. Here the quarters are bent, the forehand following a straight line. To perform a *travers*, keep the horse's forehand on the outside track of the school and bend his quarters inwards from the wall or fence, again giving a three-path movement. *Renvers*, keeping the quarters in towards the wall or fence and make a track 3 to 4 ft from it. These movements are supposed to bend and supple the horse's spine, but I think they are of more use to the rider than to the horse as they teach him to have more control over the movement of his mount. I do not like them much myself and will quote from Henry Wynmalen in *Equitation*, "This type of movement should never be ridden but for very short distances at a time. The combination of rein and leg aids which is needed to bring about these 'flexed positions' is indeed rather complicated and requires expert horsemanship without which it is not advisable to attempt this particular work." Should the rider feel that his horse needs more suppling in the back I would suggest that he works his horse on circles, increasing and decreasing their size and thus getting 'shoulder yielding' to start with and eventually a complete half pass both ways, as the horse becomes more obedient to leg and hand and the rider more adept in his control of the body of the horse. On page 25 I have described how to do this work and how to ride a correct circle with the aid of sawdust. The same idea can be followed very easily by the use of a rake in a covered school. The use of markers is difficult since they get in the way of the horse.

Fig 84. Mrs Lorna Johnstone on El Farruco, *member of the British Olympic dressage team at Munich, 1972*

Next one comes to the half pass, i.e. the horse moving forwards and sideways, crossing his legs in front and behind as he goes forward. This is a beautiful and graceful movement when correctly carried out, but an awful affair when the rider and/or horse is wrongly bent or is resisting. The horse must be flexed and, I think, collected to some degree, and looking in the direction of the movement. It is easier to work at the trot than at the walk as impulsion is very easily lost. Put the horse at 45° from the wall or fence on coming from a corner at a sitting trot, leaving enough room to go forwards on to the track, and ask for a few steps sideways. Then push forwards into the rising trot, patting the horse if he has tried to understand this new movement. Here again, if the rider has ridden a well-schooled horse to get his aids correct it does help him enormously to get the feel. I never like two novices together, they bother and worry each other. The old-fashioned way of pushing a horse over in hand into a stick while on the ground did give him an idea of the movement sideways, but I have always found the forward impulse difficult to maintain. One doesn't expect the rider to find movement in two directions at the same time immediately simple.

The *rein-back*, from a rider's point of view, is a good movement in order to learn the feel of the horse dropping his head in acceptance of the bit and his readiness to obey an order. Having halted the horse, the rider asks gently with his hand, helped by a gentle pressure of both legs slightly behind the girth, that he shall accept the bit and drop his head. When he feels this, he can ask for the rein-back. The horse moves backwards, held straight by the rider's legs. The horse should be in two-time, i.e. moving diagonally as in the trot. Only a few steps should be asked at a time and the horse should then be put straight into a forward movement so that he neither gets behind his bit nor develops a run back instead of a step backwards.

The *counter-canter* is a suppling exercise for the horse and also proves his complete obedience to the rider's aids. The horse has already learned to canter on the inside leg of a circle and now the rider is going to ask him to canter with the outside leg leading. To start this work, the horse should be asked to canter on large serpentines holding the same leg. At first any horse who is well-schooled will try to change legs automatically as he changes direction; but the rider must hold the horse on the same leading leg as he goes through his bends–in time the horse will begin to understand what is required of him. The rider must have patience and realise the horse is having to un-learn his previous training, and that leading with the outside leg is more uncomfortable for him, only large circles should be asked for. In the end, the horse should be able to perform figures of eight at the counter canter. I do not care much for this exercise for show horses or for cross country horses as I feel it contradicts their previous training and, having spent hours instilling natural balance, it seems a pity to muddle the horse's brain. It is also inclined to make a horse feel he has to rely entirely on his rider, which I feel is wrong for the cross country horse who should be able to look after both himself and his rider when in trouble. But I do agree it is a test of the rider's aids and the obedience of his horse, and I realise that the B.H.S. Test 18 is used for event horses.

Cantering from the halt or *rein-back*. This is asking a bit more than simple transition from one pace to another, up and down. It can be likened to a car going from the halt to top gear, or even from reverse to top gear when doing the rein-back to canter. Some horses really seem to delight in this exercise but the difficulty is to keep them calm and their movements smooth, and not to let them anticipate the aids of the rider.

The most spectacular of all movements at the canter is, I feel, changing legs on a straight line. So seldom does one see this done correctly and smoothly. The horse needs to be able to be collected and must be very responsive. The rider must sit very still and keep the horse on a straight line. So often there is too much movement from the rider and his body, which swings the horse from one side to the other and looks most inelegant.

Two-track work

This must be done carefully so that when the horse is on two tracks he is always moving forwards and his legs cross in front and not behind. A lot of people prepare the horse for this work by making him move away from the rider's legs on shoulder in, or shoulder out, where the horse is making three definite tracks with his legs. In other words, should he be doing shoulder in towards the right, leaving a hedge or the wall of a school on his left, his head should be flexed to the right, his off fore will make a track of its own while the off hind will fall into the track of the near fore, leaving the near hind also to make a track of its own. The horse's legs are therefore moving in a three track movement. This exercise should be practised on both reins. As another exercise to increase his control of the horse's hind quarters, and also to work on the sensitivity of the horse's sides, the rider can practise the pivot on the forehand. Here the horse should pivot on one fore leg, crossing one hind leg in front of the other–it is permissible to take one step forward, but the horse should never move backwards. Pivot on the quarters can also be practised to gain more control of the horse's forehand. At first when asking for two track movement the rider should be satisfied with two or three steps correctly executed. It is easiest to place the horse a little distance from a fence, or wall of a covered school, at an angle of 45 degrees and ask him to walk both sideways and forwards; before reaching the object then turn the horse large and trot away to maintain forward movement. After initial practice has been done, preferably on a horse who knows his work, it will be found easier to execute this movement at the trot rather than the walk, because at the trot the horse has more natural impulsion.

Fig 85. Place the horse a little distance from a fence

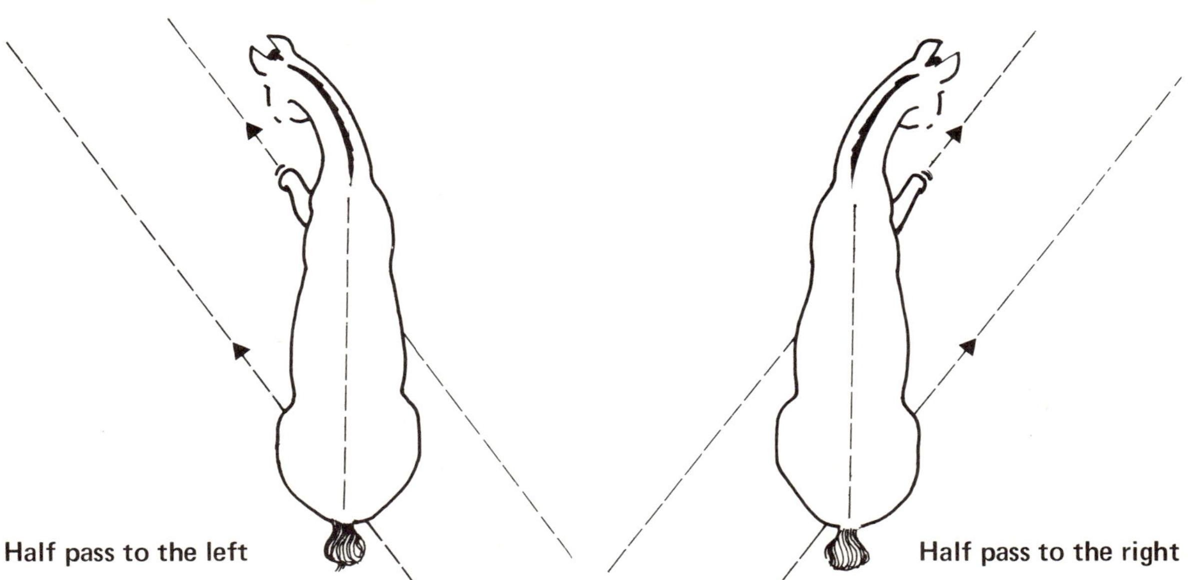

Fig 86. Diagram to show half pass to the left and right

SIDE SADDLE

Fig 87. A small light rider can control a big horse with a much steadier hand

There has been great interest shown recently in side-saddle riding and I am adding a chapter on this type of equitation since I am one of the few left who really rode and hunted side-saddle in the past.

When I was 15, my father told me, 'No daughter of mine will ride off the farm cross-saddle.' In those days one did not argue with one's parents and, to help matters, I adored my father so I only wished to please him. He was, and still is, an image in my life, one of God's natural horsemen: he was one of England's first riders to compete abroad. He won the International High Jump in 1909, the Queen of Belgium's gold cup for International cross-country competitions in 1903 and 1904, besides winning an untold number of races, point-to-point and hunting both our family pack and, sometimes, the adjacent carted stag-hounds.

I am assuming that the rider can already ride cross-saddle well enough to guide her horse. But even so, the instructor should provide a sensible, quiet horse to teach on. First of all, the rider must mount: a leg up is the easiest way of mounting. Not the old-fashioned way of stepping into your groom's hand, or that of some young gallant who would dirty his hand for you, but a leg up with an ankle push. The push must be rather more forceful than that needed to put a lady into a cross-saddle, because ultimately she must arrive up on the saddle so that both her legs are on the near side and she can swing into the pommels straight away.

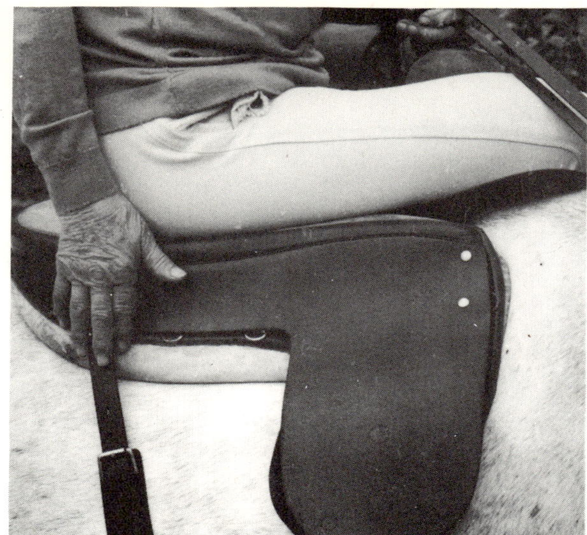

Fig 88. The right hand on the balance strap

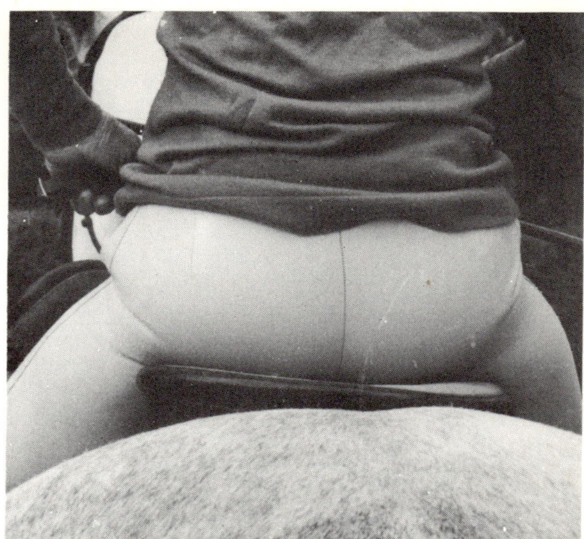

Fig 89. The centre seam of the breeches should be vertical above the channel of the saddle

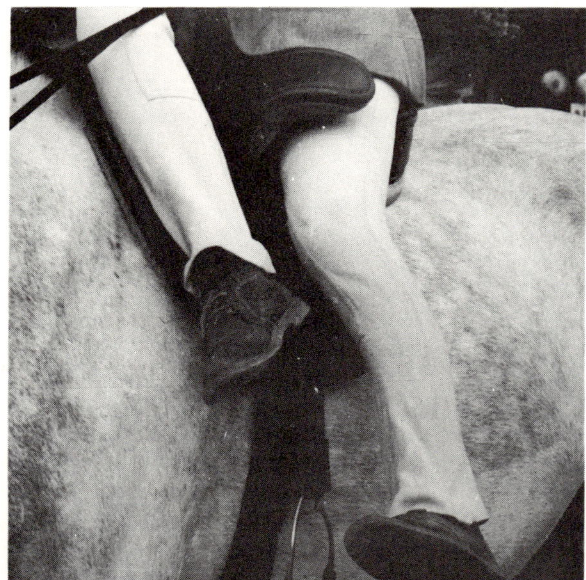

Fig 90. The right toe is pointed down towards the left toe, which should be pointed upwards

To start with I would explain the three ways of mounting. First I would recommend the ordinary way of getting up cross-saddle with one leg on each side, because from this position she can swing her right leg over the saddle easily and put it round the top pommel. Secondly, the usual way for a lady to be legged up by a person who is experienced—and there is, I can assure you, quite an art in doing this correctly so that the skirt, kept well out of the way, can be arranged over the right leg when it is in position on the top of the pommel and the left leg slips into position under the lower pommel. Finally, one can, in fact, mount quite easily unassisted; to do this, face towards the horse's head and, taking the reins in the left hand, put the left foot into the stirrup, being careful to take the weight on the outside of the iron so that the knee can bend away from the horse to allow the right leg to pass between it and the saddle. Again one has to be careful to keep the skirt well tucked away to the left.

When the rider first sits on a side-saddle with her legs in a cross-saddle position, she should be careful to sit in the centre of her saddle and should then put her right hand on to the balance strap, keeping her shoulders square and looking forward through the horse's ears. Now she should swing her right leg over and round the top pommel, still keeping her body straight. In most cases she will feel a slight twist above her hips. The centre seam of her breeches should be vertical above the channel of the saddle. Taking her reins in her left hand, she should start to ride with her right fingers still touching the balance strap, being careful to keep her shoulders level, and this will help her to settle herself into the saddle. It will be found by the new side-saddle rider that the most comfortable paces are the walk and the canter. After a short time the instructor should correct the position of the legs on the near side. It should be explained to the rider that if her right toe is pointed downwards towards her left toe, which should be pointed upwards, this position will give her a firmer seat, since the downward thrust of the right toe pulls her knee downwards into her top pommel, and also the pointing of the right toe will give a good line of a skirt when riding. The left knee should be placed comfortably against the saddle and the rider should

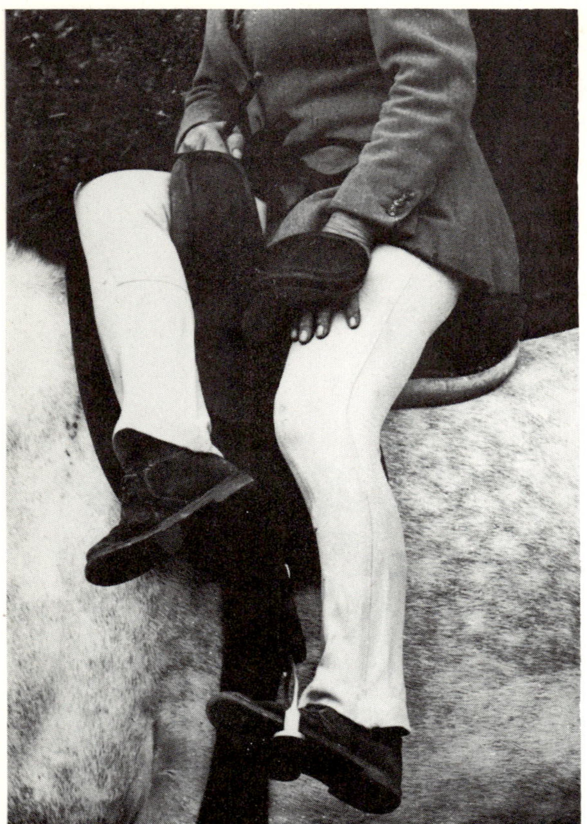

Fig 91. The rider is able to pass her left hand between the left thigh and the under pommel

be able to pass her hand between the left thigh and the under pommel. To start with, I think it an excellent idea that no side-saddle rider should use a stirrup for at least two or three lessons so that she does not have the temptation of putting too much weight or reliance in her stirrup, thus pulling her saddle to the near side and tending to come off the top. When the stirrup is used it should only be as a foot rest (I was made to ride often for four hours on end without a stirrup when I was learning to ride side-saddle—at least it made me sit on the top of my saddle and my stirrup has never been of any importance to me—but that was the hard old school).

Some horses are much more comfortable side-saddle than others. A slow, controlled trot (I did not say a collected trot and this pace is not talked about except in advanced dressage) is a pace which the schooled horse can do and is found comfortable both cross- and side-saddle when the rider is asked for a sitting trot. I would never suggest that a side-saddle rider be taught to rise in the early stages of her riding. In days gone by a lady's side-saddle horse was especially trained. The training consisted of its being light in hand and being able to perform a comfortable slow canter always on the off fore, as cantering on the near fore is not so comfortable for the lady as it jolts the rider into her pommels. The idea was that, while hacking, the lady's horse could canter slowly on the verge of a road while her escort could trot along beside her. Such a horse was kept merely for side-saddle work and valued for its ability to give only the off fore in cantering. So much for the old approach: today that much-valued lady's hack would be termed one-sided, ill-schooled and would be discarded as needing re-schooling!

At first, the learner must ride a good deal under supervision. I say 'supervision' since someone must watch to see that the rider sits straight in the saddle with her shoulders square and does not collapse her left hip. A lady riding side-saddle must look elegant, sitting upright with her head held high. Nothing looks worse than someone who pokes her head forward—imagine this forward head carriage with a top hat on it.

There are few tailors, if any, left who know how to make a pair of lady's side-saddle breeches. These should fit very tightly and be smooth under the right knee when bent round the pommel. Therefore today most people wear their cross-saddle breeches under their dark skirts and it has become customary to see these light-coloured breeches showing, but in the olden days all ladies had breeches made of the same material as their habits. Should the rider have trouble with a rubbed right knee because of the wrinkles in her cross-saddle breeches, I would suggest that she put a small piece of sponge under her knee to

Fig. 92. A lady on a side saddle has a much firmer & more independent seat

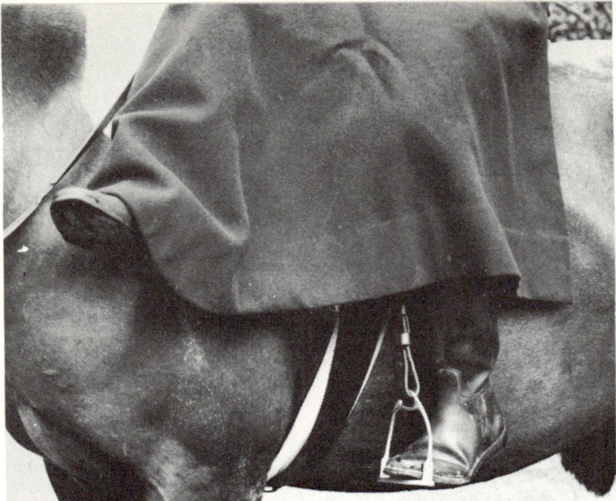

Fig 93. The tailor's headache–and incorrect positions of both feet

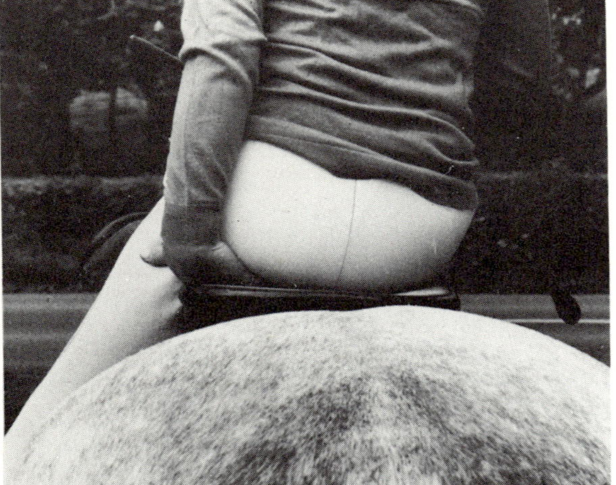

Fig 94. Showing the rider's weight on the right seat bone

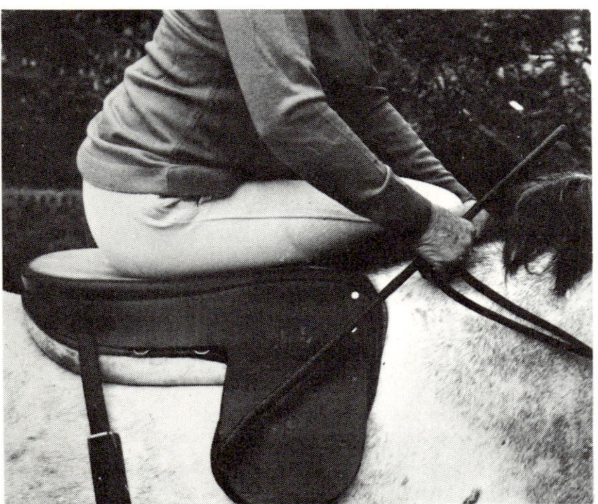

Fig 95. The forward position for jumping–weight on to the right knee

take the wrinkles away. When I first hunted I often had sore knees because my breeches did not fit. Great was the day when I had my own habit made for me! As in all things, fashions change–in the early 1920s, no lady showed her left foot, the skirt covering her boot by at least 4 in. Now the more of the left boot that can be seen, the more fashionable the rider. This long leg effect does tend, in my mind, to make the rider slip off her saddle towards the near side unless she uses a very firm knee grip with her left knee to keep her body up on top of the saddle. I feel that this rather exaggerated position should only be used when showing (i.e. the rider does not ride for long periods) but a shorter stirrup and leg position helps the rider to stay on top of her saddle and have a stronger seat, which is more practical for hunting and long rides. One often sees the younger generation riding side-saddle with the right toe stuck out beneath their skirts which makes the line of the skirt most unattractive, in fact, what I call a tailor's headache.

When considering the position of the lady in the side-saddle, it must be stressed that the whole of the weight is taken on the right seat bone and thigh and the closer the right thigh is pressed on to the outside of the right of the side-saddle, the better. One should be able to pass a hand underneath the left seat-bone of any balanced side-saddle rider. It must be admitted that perhaps this is an unnatural way to ride, since the lady's body is slightly twisted to the right, but with pommels the rider feels very firm in her saddle and I have found that horses which buck are much easier to sit on riding side-saddle than astride. A small, light rider is so much stronger on a side-saddle as she can sit against her pommel and control her horse with a much steadier hand. The old idea of a lady's hands always being better than a man's was, I am sure, only because a lady on a side-saddle had a much firmer and more independent seat and her hands were therefore steadier. Another advantage of riding side-saddle is that a small, light rider has better control of a big, strong horse, besides looking much more in keeping riding a heavyweight horse. Both my daughter and I have had great fun when riding huge horses in the show ring and do look so much less like tom tits on gateposts!

When a lady starts jumping, she should have a forward position the same as required for cross-saddle jumping. She will then find her jumping easy and comfortable.

Fig 96. Family pair – The author and her daughter 'Jinks' riding Chocolate Box *and* Tara

SCHOOLING THE HORSE

INTRODUCTION

The first principle of schooling or breaking is establishing an understanding and trust between the human and the horse—whatever age the latter may be. Once confidence is gained it is so much easier to progress with teaching the horse to accept the unaccustomed things we demand from him and, when the necessity arises, for the veterinary surgeon to give injections and cause bodily discomfort. If the horse is held by the person he knows, who can talk quietly to him and pat him to reassure him, he will then be much more likely to submit calmly to treatment.

This first approach matters so much because the horse is a creature in whom fear is the predominant factor; consequently, he is always apprehensive. In his wild state he is given to dashing away from anything he does not understand, or that frightens him. It is best to approach any strange horse by speaking quietly but reassuringly to him. We often come across horses who have been mishandled, or have not been handled at all—principally our native Mountain and Moorland ponies, born in the wild state, rounded up at weaning time, parted from their dams, and sent to

sales. Is it any wonder that their first impression of the human race is one of fear, connected with leaving the comparative safety of life with a mother in familiar surroundings, and entering upon the frightening experience of travelling to and from a sale, eventually to arrive, perhaps alone, in strange surroundings? It speaks well for the inbred temperament of most of our native ponies, and the toughness of their make-up, that they come to hand so well, and forgive such abuses in infancy. Any with funny temperaments can hardly be blamed for their mistrust of those two-legged animals who sometimes behave so thoughtlessly towards them.

Animals bred on private studs are usually more fortunate in that they are handled from birth, taught to lead, have their feet attended to, etc. When travelled they are carefully led, and have someone they know at their heads on a first journey.

EARLY TRAINING

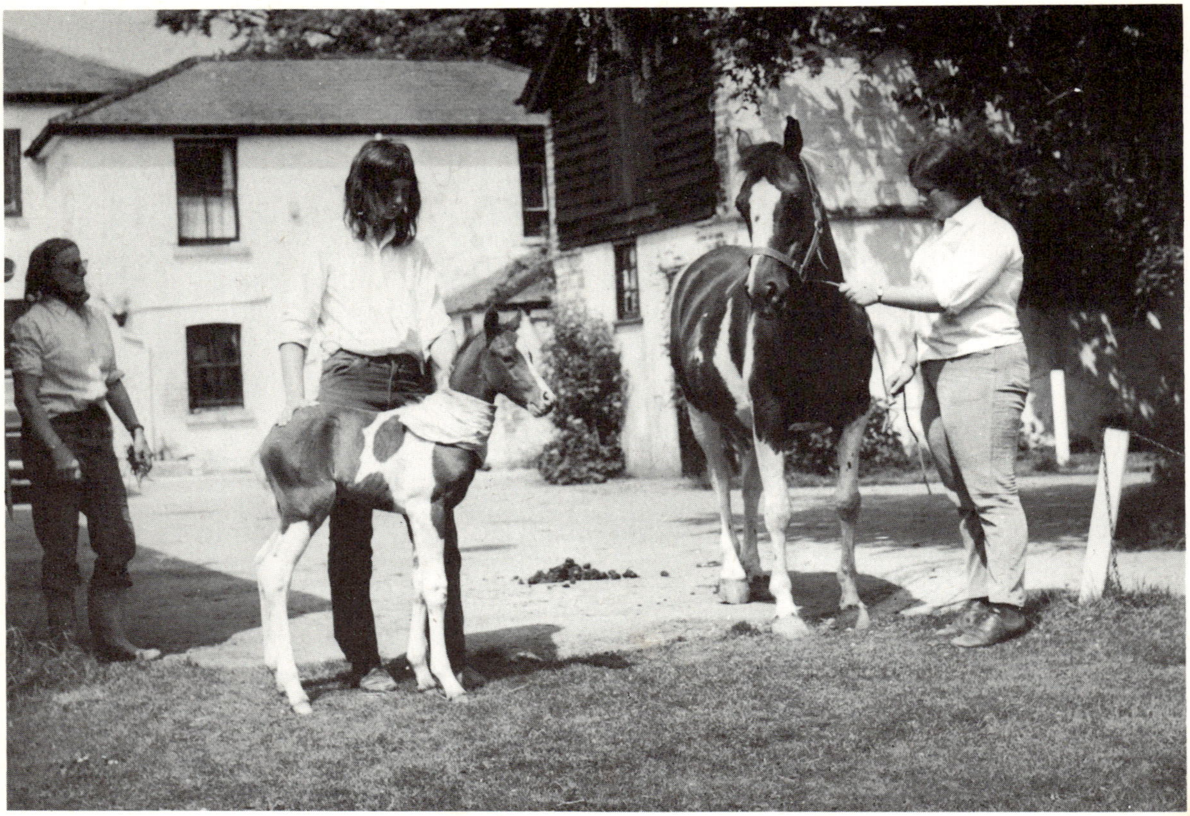

Fig 97. A few days old

I have often been asked how soon a horse's training should start. I feel the answer is, as soon as possible if you are lucky enough to breed or train your own horses. The first, most important thing is to gain the confidence of the animal, however small. A foal is always most curious, but also afraid by nature, especially of something taller than himself, so if one stoops to the level of the small creature he has more confidence to investigate the human being whom he will have to serve in return for being cared for.

The first real training begins about 48 hours after birth when a foal is led out with his dam, if he is born in a stable, or when he is handled in the field. All handling of foals is greatly helped if its mother has confidence in the person who is handling the baby. Some mothers are so possessive that it is very difficult to touch the foal for a day or two, in which case it is best to leave the first handling until the mother has settled down. Or, if you have been able to catch the dam, one person can hold her, if possible with a feed, to take her attention off any approach to her precious offspring.

Fig 98. Someone smaller than himself!

After a few days a foal slip can be put on the

Fig 99. Haltering is easy when the foal has been handled from birth

foal's head and with one hand behind its quarters it can be led out to the field daily. It is surprising how soon the foal will follow his mother and learn to lead properly. It is all-important that the leader of the mare be observant and see that the person leading the foal is not left too far behind. If the foal loses sight of his mother round a corner, he may just stop with all four feet planted firmly on the ground; then the leader of the mother must retrace her steps and fetch the unwilling child.

Haltering

Haltering a youngster for the first time can be very difficult if, for one reason or another, the animal has never been handled from birth. There are many ways of doing this, some very drastic. It may be a foal that has run with its dam since birth and you are about to wean it, or it may be a purchase at a Mountain or Moorland breed sale, when young stock have been running wild on unfenced common land. It is usually fairly easy to get the youngster into a loosebox from the lorry in which it has arrived, given enough people to line the route from one to the other. Having shut the top door as well as the bottom, leave well alone until the animal has had time to settle down and find its hay and water. It helps if, in the adjoining box, you have a quiet horse, and there is a window or bars through which the stranger can watch you handle and feed the other horse. Slowly you can approach the newcomer, and just stand and talk, offering a nice piece of grass, a delicacy he has not seen for some hours. If the animal is confident, he will take the grass, and should he look for more your battle is half over. In a matter of days you will be able to handle the head and neck, and put on a head collar. If you are not so lucky, the animal will shy away from you, go to the furthest corner, not look at you, and refuse to go near you when you come to feed him. Remember, then, that the animal has probably never tasted corn, and will not understand a short feed of oats offered to tempt him.

This reminds me of a funny story against myself. Some time ago the police awoke me at about 3 a.m. saying some horses were on the road about two miles

Fig 100. A rope attached to a stick . . .

Fig 101. . . . is manoeuvred into place

Fig 102. The rope is knotted to the halter

away—would I come in case they were mine? Getting some whole oats in a tin bucket to rattle well, I got into the police car, armed with headcollars plus bucket. When we reached the 'scene of the crime', there were three large, dark forms. I felt sure they were not shaped like mine! When I rattled the bucket the 'divil' himself might have appeared; the horses took to their heels and disappeared at full speed. They definitely were not mine.

The usual way for most people to halter an unhandled animal is with a crooked stick. After some careful manoeuvring, the halter goes into place, and you can leave your halter on with a rope hanging, which is useful to ease the animal near you. The rope should be knotted to the halter, so that if the pony treads on the hanging rope he does not hurt his jaw. This hanging rope can prove very useful to catch hold of while the youngster is still afraid of you. Then, by having the neck and face gently rubbed, he will soon submit to handling. Proceed each time carefully a little further backwards before finishing. If you can go to the horse at least 5 or 6 times per day at first you will make great progress. In the old days, this was called gentling—which it is. I remember once having a wild mare, 7 years old, sent to me. I gentled her for 3 days—you couldn't touch her head (she had had a very badly split ear) and her heels were very handy defence weapons. It took about 2 to 3 hours per day, always talking and stroking, and then I could handle her all over and pick up her feet. Mind you, I worked hard, neglecting everything else, goaded by my daughter who felt it was a waste of time. In the end she proved right—the mare was never reliable—fear was always uppermost in her makeup.

Another harder way is to keep water from the animal and only offer it in a bucket. Allow a small drink, take it away again and keep returning so that the animal gets into the habit of coming to you for what he wants. Soon you will be able to touch him, and go on from there. Having handled his head well, always rubbing behind his ears as he drinks, slip the noseband of a headcollar into the bucket, scratching his ears as usual. Then carefully slip the head strap over the head with the other hand, while someone else takes the weight of the bucket.

The last and most drastic method is to rope the horse. Slip a slip knot over his head with a stick as you would for a halter; the terrified animal will plunge and fight and eventually go down. This is when you quickly slip a headcollar on (leaving a long rope attached to it so that you can haul the horse towards you later), loosen the rope round his throat and let him get up. I hate to see this method used as it is very dangerous, since the horse can badly hurt both himself and anyone in his way. It should only be used as a last resort when all else has failed. I have never known the water trick to fail myself. This last method I only use when I have had to halter a youngster in one afternoon and cannot spend the time it should normally take.

You can now proceed to gentle your horse and break as one would normally do. Unhandled horses are slow to start with but usually come to hand very quickly once one has gained their confidence. The old horse-breakers of days gone by always said you had to have one good fight and the sooner that was over the better. Then the horse became the servant of man for ever. But there are many I can think of that I have never known to fight at all.

Drastic measures should only be used as a last resort—the quieter you can be the better, but we all know there comes a time when human patience runs out. Your problem may be a foal or a nine-year-old unbroken animal; usually the older they are, the more set in their ways they will be and more of a problem they will prove.

Some old books suggest that the trainer throw the horse daily to make him realise the power of man, his master. I have never practised this, though I have several times accidentally thrown a horse on the lunge when he got tied up in the lunge reins by mistake. It is usually a very difficult one who fights and rears badly, getting his legs over the reins. I believe in never letting go, if possible, since the horse then learns he is boss, and that every time he pulls away he has won that round of the contest.

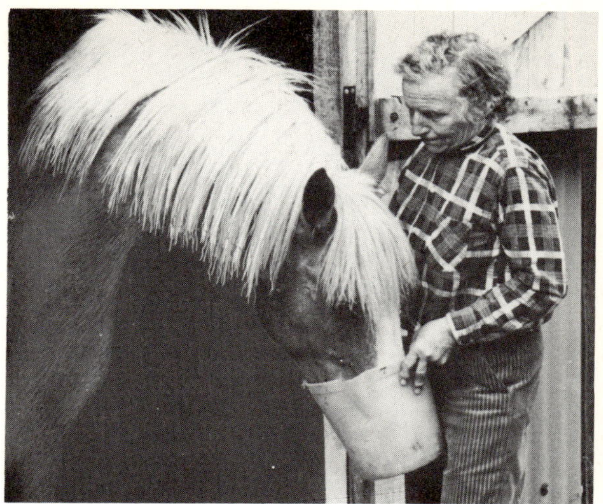

Fig 103. Offering water in a bucket

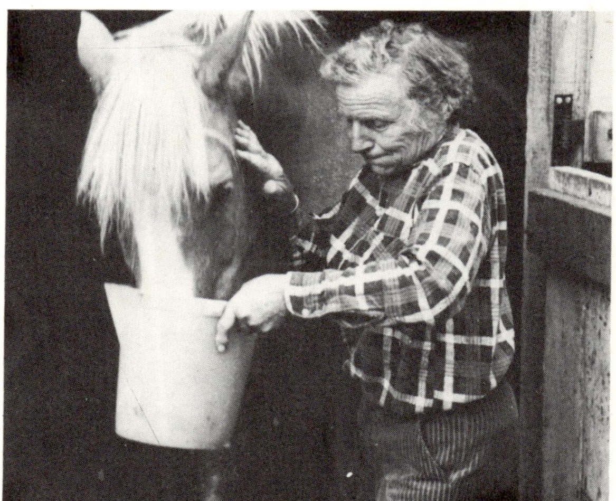

Fig 104. Soon you will be able to touch him

Fig 105. Put the headcollar in the bucket

We had a 9-year-old mare given to us who was quiet enough to handle but had never been broken. She had a completely confident outlook on life, but also a completely obstinate one; why should she do anything she did not want to do? Why start being obedient to man's wishes at 9 years old? She had lived well in a field for those nine years, food was provided when grass grew short, so man had always been her servant–why reverse that pattern of life now? She would not lunge, unless led; she fought and she plunged when the whip was used to push her forward, she got herself, me and my helper well and truly mixed up. Down she went, roaring with rage, and she lay there, unable to get up. I let her lie there and my daughter, who was schooling in the next field, came over telling me I would kill the pony (she was not being very helpful because she was frightened, never having seen such a performance before!) After a short time the mare gave in and lay quite quietly, breathing heavily, I knew round one was mine. Talking to her and patting her, I very slowly untwined the lunge rein from her legs and let her up. She had not even grazed herself. She stood for a few moments and from that time onward I had no trouble. To this day (I still have the pony) she is a beautiful ride and one of the most perfect to teach with on the lunge. She is always in good condition in the winter (though inclined to get too fat during the summer), a really good tough pony who hunts, jumps, gymkhanas and to whom nothing comes amiss. I am very grateful for the gift of an unwanted, unbroken 9-year-old mare, a nice useful member of my four-legged family, but her one fault is that she is inclined to kick other horses and she has never got over her hate of being confined in a box–she has to be tied up in the yard as she frets and sweats in a stable–but travels well in a lorry or trailer. This love of freedom must stem from her 9 years in the fields.

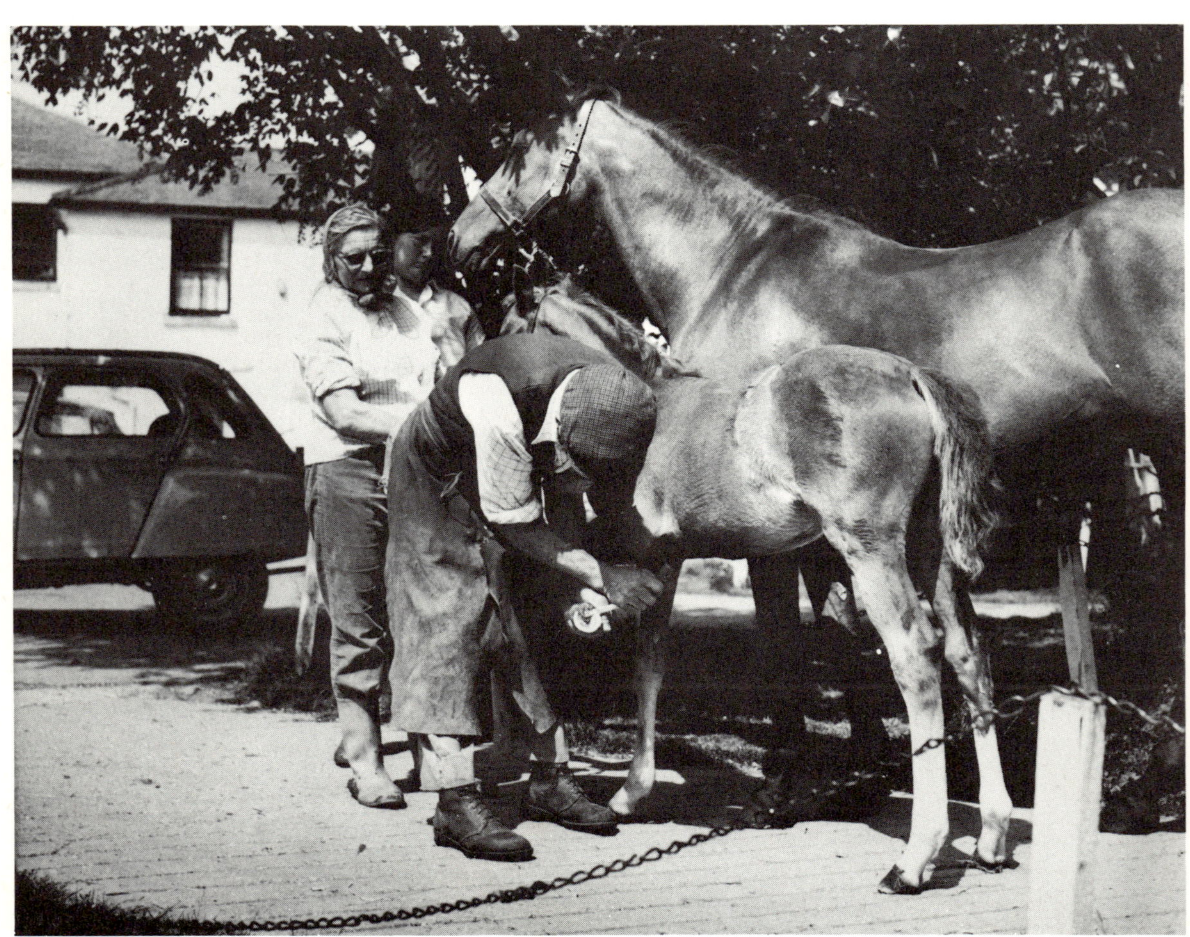

Fig 106. The blacksmith's first inspection–always stand the mother close by to give moral support

The foal born on a stud has now learned to be handled all over and to be brushed very gently, but if there is any difficulty, a good approach is to scratch its bottom as this attention always gives great pleasure—it is, I am sure, just the same as we ourselves really enjoy a good back-scratching at times. The feet should be quietly picked up and handled and picked out with a hoof-pick so that the foal is not surprised or frightened at its first introduction to a blacksmith. Always stand the mother close by to give moral support when strangers are about.

Now life at the stud or home goes on much the same every day—gentle handling, firmness but kindness, with the horse learning discipline but having a wonderful time in the fields in the summer sunshine until the day comes that it is decided that the foal is to be weaned i.e. about 7 months after birth.

If possible, it is always advisable to take 2 or more foals away together so that they are company for each other and their dams also have company, as it is obvious that dam and foal would both miss each other (some much more than others). There are many ways of parting the mare and foal. Some people separate them for an hour or two at a time and gradually wean them. Then there is the method used by large studs of taking one or two mares out of a field of ten mares and foals and each day taking more mares, leaving only the very kind ones to the end. This method I have never tried and I doubt if I could, as one can never make a scheme work unless one has faith in it: for instance, some foals could get into great trouble trying to get to their dams on seeing them leave the field, or trying to get too close to another mare when thirsty and lonely. The most usual way of weaning is to lead the mare into the box in which you are going to wean the foals then slip each mare out, leaving the foals in a safe box, and take the mares out of calling distance so that they cannot hear each other. Do make sure the foals are in boxes where they can see out, rub noses with each other, yet have no temptation or possibility of jumping out.

I had two boxes especially constructed with bars half-way along the front and some quarter of the way between the boxes, which had extra high doors with tops as well. Put a good bed down with plenty of sawdust on the floor covered with straw so that, however excited the foal becomes, he cannot slip and hurt himself if he dashes about the box. Some get quite hysterical, but usually with a nice feed in the manger, a hay-net hung high and a bucket of water tied to the wall he will settle down.

Some people advocate weaning two or more foals in one box—this I have tried several times without

Fig 107. Donna—*the mare who was broken at the age of 9*

Fig 108. A safe box

success. Sometimes the youngster will get so fussed that he kicks his friend unmercifully, even though these two foals have been great friends turned in the same field since birth—this is just pent-up distress. The other objection, in my mind, is that one is always boss and pushes the other off his food etc. I like to know how each foal is eating and what his droppings are like when weaning, as some can fret so much that they go off their food and become very constipated. After the mare has been taken away for an hour or two, the foal should be groomed by someone he knows. After about 24 hours, turn the foals out together in a familiar field but ensure that they cannot see or hear their dams. The thing that has always amazed and saddened me is that the dam usually dislikes its yearling or 2-year-old but that young stock turned out together seem to form a life-long friendship. After some years mother, daughter and grand-daughter will be very happy brood mares together. We have one dear old mare who mothers all the weanlings in the autumn after they are weaned—and protects them should another horse be turned out or get out into her field.

For the first winter, all young stock should be kept in at night and turned out by day.

So the first Christmas comes and goes—the snow is a source of great amusement and fun to the young ones. Hard weather is dangerous unless the foals are turned out regularly for a short time and do not rush about. I often use a straw ring, fenced, for exercise, as I am terrified of a bad fall marring the future of some animal and feel a lot of those 'backs' are caused by a youngster falling awkwardly.

The spring follows and by May the yearling should be out for summer grazing and enjoying the good grass and sunshine. Should the youngster be shown in hand, then it will be brought in and handled daily, groomed and led in-hand for a week before the show. It is as well to try and bring a friend into the next door box as so often a lonely yearling will fret for its friends and go off its feed and lose condition.

Showing in hand, if properly done, is all good education for the future. He must learn to lead well, to box and stand quietly when tied up, and the experience of travelling to shows, seeing fresh places, crowds of people and horses, and still remain calm, is good for him.

Daily visits to the field in the summer evenings

Fig 109. Showing in-hand

are a joy to all concerned and so the summer passes peacefully and the young stock grow. The following autumn a plan must be made as to what is to be done during the coming winter–Mountain and Moorland yearling ponies can winter out very well if regularly and adequately fed. Some better-bred animals, half Thoroughbred or Arab, should come in at night so that they can sleep in the shelter of a stable and get two good feeds a day. The same applies to the 2- or 3-year-old. Regular attention to feet and worming is also important.

Some people who have good sheltered fields or have field shelters do keep their breeding stock out all the year round. This certainly reduces work and breeds youngsters that are hardy. The only important point is to watch that each animal is getting its fair share of supplementary food, i.e. no one horse must be so bullied as to lose condition. It is certain that one does not save on the feed bill as horses eat more, turned out in winter, than those kept warm in stables, since half the food eaten goes to keep them warm and the other half to fulfill their bodily needs. Besides, food is more easily wasted when fed in racks and trodden underfoot. Company and good feeding produce contented stock that do well and are a credit to their breeders.

When the young horse is 3 years old, proper training can be started. Some people advocate breaking at 3 years old. Yes, if you want to pass on the youngster, you are more likely to be able to sell it if it is quiet to ride–but if you want to keep it, it is better to part-break it in the spring and turn it away again for the summer and autumn so that no horse is really worked before it is 4 years old. Even so, one often finds that certain animals are backward and need yet another year to be able to stand up to being worked. They are like human beings–some mature quicker than others and some out-grow their strength.

Fig 110. Daily visits to the field are a joy to all concerned

PART BREAKING

Fig 111. First lessons on the lunge–with whip held high and an assistant at his head

By part breaking, I mean teaching the horse to lunge quietly and also to long rein (optional), to carry a saddle, and to be quiet to mount and dismount, and to carry a person calmly at the walk. It is very necessary to give a horse time for its bones to calcify and grow strong enough to stand up to carrying the weight of the rider and balancing itself without undue strain.

Lungeing a horse is the first training given to any horse or pony, be it intended for riding or driving. It is the teaching of obedience to the voice, the rein and the trainer's wishes.

At first, a cavesson, a lunge rein and a long whip are needed. Some people only use a headcollar to lunge with, but if a thing is worth doing, it is worth doing properly. It is best to equip yourself with a cavesson which is easily adjustable and fits most noses. The adjustability is in the length of the headstrap, cheek-strap and chinstrap. So many cavessons are very impractical since the people who make them are not familiar with their use. There are many different kinds, and one has to choose a serviceable one–not too light so that it is not strong enough to stand up to a youngster who fights the resistance of the lunge rein, yet not so heavy and cumbersome that its weight is uncomfortable to his head.

At first, when the youngster is not using a bridle, adjust the noseband to just above the corners of the mouth as the lower it is placed on the nose without interfering with the breathing, the easier it is to control the animal on the lunge rein. It is a great help to have an assistant to lead the animal forward so that it gets the idea of walking round the trainer on the lunge and stopping at the word of command, which can be 'Whoa' or 'Halt', if you prefer the German word to the English.

When the horse will walk, trot and stop with the aid of the assistant at its head, then try to get it to go for you in a corner of the field or covered school, so that you have two sides of the fence or walls to work against. Gradually, as you get the animal lungeing better, you can increase the size of the circle until you can go in the middle of the field. The position of the trainer and the use of his whip is all-important.

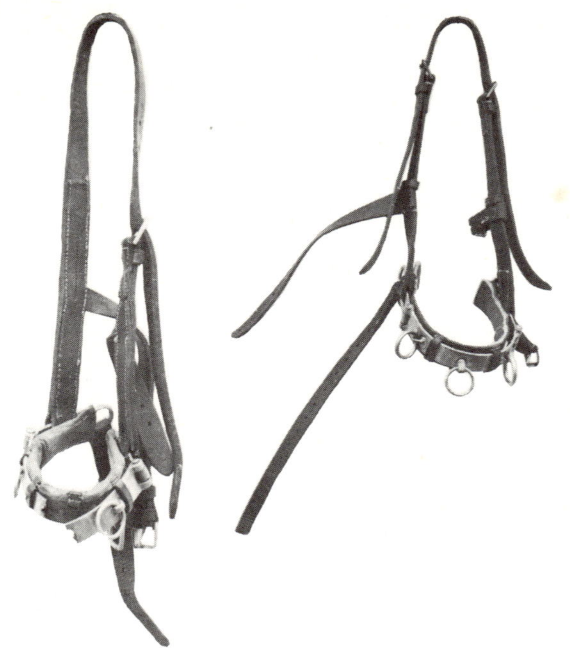

Fig 112. Three ring cavesson with plenty of adjustment, but the one on the left is too light and not strong enough

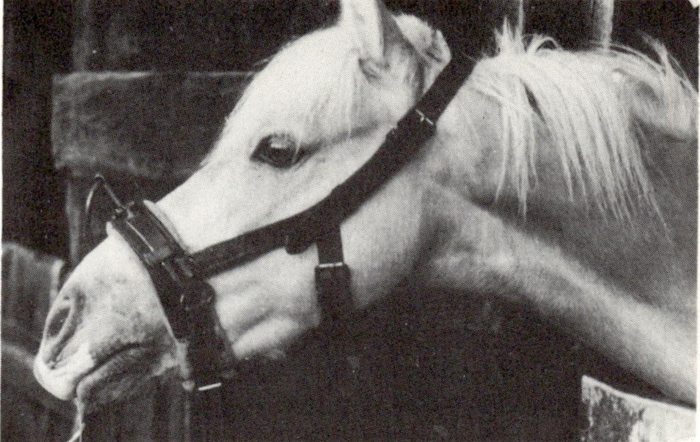

Fig 113. Cavesson on 12-hand pony

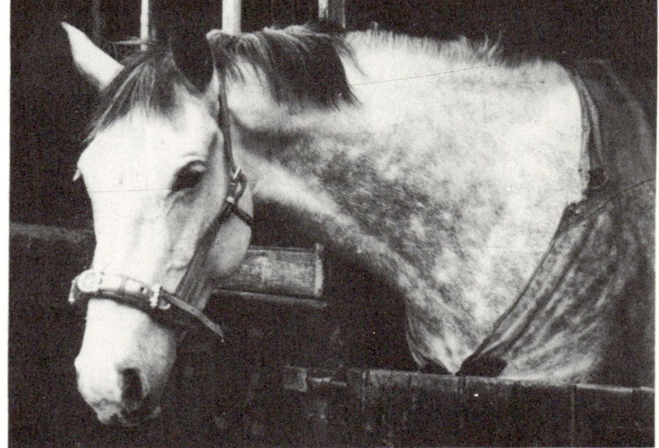

Fig 114. Same cavesson on 16.3-hand hunter

Fig 115. Using two sides of a fence

The horse should respect the voice and whip, but never be afraid of either. The voice should be encouraging and authoritative: the whip positions the horse on the circle and follows the quarters to keep the pace, but, should the horse come in towards the trainer cutting into the circle, the whip pointed to the shoulder will keep the horse out on the outside of the circle. In this case, the whip should be shaken at the horse's shoulder, very slightly, or given a slight flick—but whips trailing on the ground are inactive and useless. The whip should be held high in the hand so it can be easily flicked on the tail if necessary, to keep the pace. Correction by voice is usually enough for most young horses: tone is understood by animals far better than words. Certain simple words of command, such as 'walk', 'trot', 'canter', 'whoa', 'steady', and praise—'Good boy', etc—are understood. Each trainer has his own pet words that horses seem to understand. Many foreign trainers use the word in English 'So so'; I personally often say 'thank you', which amuses most people greatly, but my horses seem to understand 'Thank you, good girl or boy'. We must laugh at our own little idiosyncrasies.

Now that the horse is lungeing well and obediently we can introduce the roller or saddle—just something that is tight round the place of the girth. Should your young horse have worn a rug in the stable, there should be no problems, but if he has not, then he will more than likely object to the feeling of a tight corset. This should be done in a field or large space, so that there is room for the horse to explore without hurting himself. Show the animal the saddle or roller, let him smell it and put it on and off his neck until he is quite happy to let you slide it into place behind his withers. Now fasten a piece of string, or, if you have one, a breast plate, round the front of his chest, so that, should you be unable to tighten the girth sufficiently before the youngster explodes, the roller cannot slip backwards into his flank and terrify him. Your assistant takes the end of the lunge rein before you move forwards, so that when the bucking starts you have no need to loose the horse and have a frightened bucking animal tearing round the field. Soon the horse accepts the new feeling and the girth can be tightened. I personally like to see a youngster show signs of a fight, as it is best to get it over than have a sulky acceptance knowing that sooner or later there must be a show-down of some sort. Of course, there are those quiet animals whom one has rugged as yearlings and who accept it all with that calm confidence which is grand to see, but even these are likely to have a quick buck or two on the lunge for the joy of living– if one did not see this from time to time I would feel the youngster was not well in himself or was not happy in his work.

Fig 116. A breaking roller

BRIDLING

The horse has now learned to carry a saddle and you can introduce his bridle.

There are a great many ideas about how one should introduce a bridle to a young horse. The mouth, to my mind, is the most delicate part of the body: the thin, sensitive membranes that cover the jaw between the front teeth and molars, where the bit lies, are the brake-linings and cannot be replaced once they are damaged. There is no repair once the sensitivity is lost and a hard mouth is made by man's inhuman handling. No horse was born with a bad mouth–bad mouths are only man-made, so great care should be taken when mouthing a horse. One must consider and ponder on many different ways and ideas. No way is perfect, so I am going to put forward my ideas and you, the reader, must decide in which method you feel most confidence.

Introducing a horse to a bit can be done in many different ways. The main thing to consider is the kind of bit you personally feel will suit the horse in question, but you must be willing to accept the fact that you might be wrong in your choice and must therefore try something else. The two major alternatives are a thick mouthpiece, usually with keys attached, or a thin mouthpiece which is usually bound with some rag dipped into either treacle or honey. This is to give the horse something to chew at which is not unacceptable to him, since the introduction of a narrow piece of steel into his sensitive mouth cannot be pleasant. There are several different kinds of breaking bits on the market, some with keys and some without. I personally advocate the use of keys as I feel it gives the horse something to play with and encourages him to use his mouth. There is an excellent bit on the market at the moment which can be unscrewed and the keys removed, should you not wish to use them, and that can also be filled with rag soaked in honey (Fig 138).

To put a bridle on a horse for the first time, I would advocate taking the browband off and having the bit only attached to one side of the cheekpiece, so that one can slip the bit into the mouth with the least difficulty, and then do up the remaining cheekpiece. The bit should be fitted so that it does not wrinkle the corners of the horse's mouth nor hang so low as to encourage him to slip his tongue over the bit (this can become an annoying bad habit). Do not leave the bit in for longer than 5 to 10 minutes on the first day but it can be put on several times during the day: I often place the bit in the horse's mouth when I take him out to lunge on a cavesson.

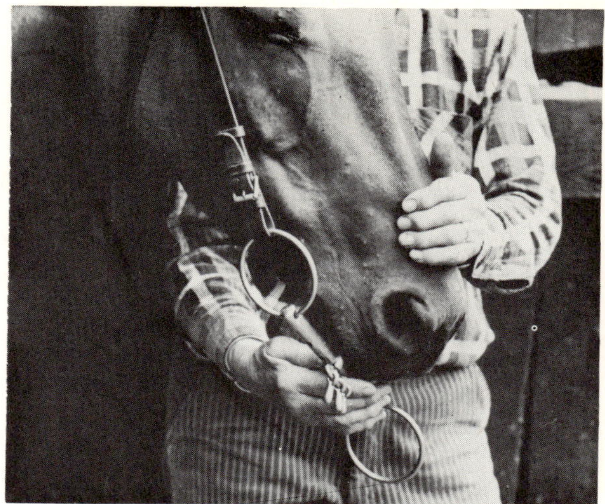

Fig 117. The headpiece is over the ears

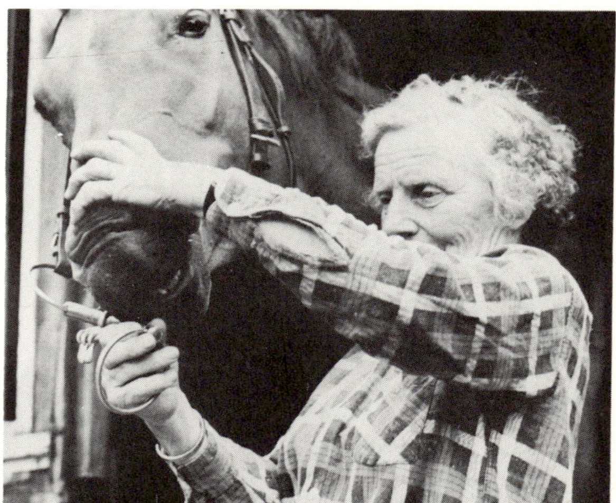

Fig 118. Open the horse's mouth

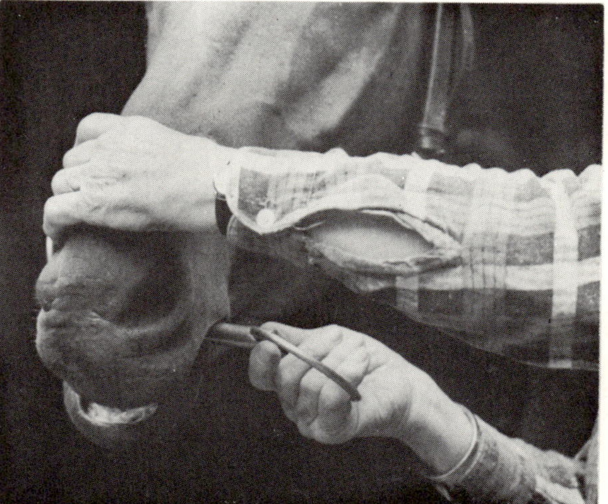

Fig 119. Slide the bit into position

After a short time, when he has become accustomed to the bit and it does not fuss him too much, the side reins can be put on the bit and back on to the saddle or roller, keeping the line from bit to saddle in more or less a straight line. Then gradually the side reins can be raised a little way each day until they are on the dees of the saddle, more or less in the same position as the rider's hands will be. From a straight bar bit, I generally introduce a half-moon snaffle. Should the horse be perfectly happy in this I often keep him in one for many months, or one can change him into a jointed egg-butt or cheek snaffle. Should the horse open his mouth and try to resist the bit, I would fit a drop noseband (Fig 137). Should he persist in trying to get his tongue over the bit, I would use a tongue-check for some time. Often after a week or two it will be found that the young horse has forgotten this annoying habit and the tongue-check is no longer necessary.

When the young horse has become accustomed to wearing the bit, we can start to use side reins to give him the feel of what hands will be like in the future. There is a choice of two kinds of side reins–fixed or elasticated. I prefer the latter since they have give in them, but some well-known trainers, such as Count Robert Orssich, only used fixed reins, as I did too at one time; it is a matter of personal opinion. Always lunge the horse without the side-reins for a while to get him settled and to get all the playfulness out of him. If he bucked with the side-reins on, he would hurt his sensitive mouth, but when he has settled down it is generally safe to use them.

Another thing to decide is the vexed question: should the inside rein be tighter than or the same length as the outside rein when lungeing on a circle? In my mind, when riding the horse you bend him on the circle with a slightly stronger feel on the inside rein, so why not do the same when one is lungeing? Many well-known people do advocate that the outside rein be the same length–to me it does not make sense, but with horses there are always two other ways of doing the same thing.

To start with, the side rein should be attached to the cavesson and be very loose. Afterwards they can be transferred to the bit. I like to put them level with the line of the bars of the mouth to the roller or girth. Should the saddle have no dees, then the side rein can be attached to the stirrup bars. Mr. Gibson, the saddler from Newmarket, was kind enough to give me a useful attachment to take the side-rein. He is an amazing old gentleman who would often show me something, ask what I thought it was for, and say, 'Well, take it home, find out and tell me.' It was he who made so many good cavessons copying the type Count Orssich used from Austria.

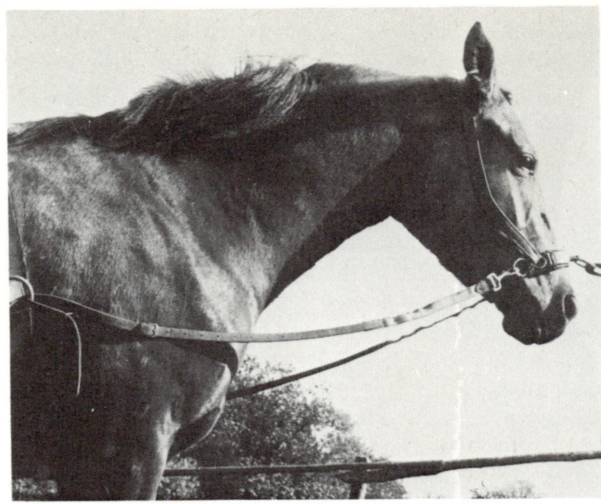

Fig 120. Elasticated side reins fixed on cavesson noseband and roller.

Fig 121. Side rein on bit and raised on girth

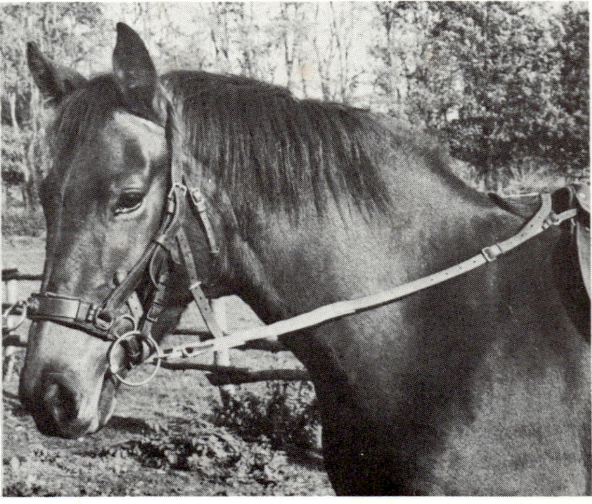

Fig 122. Side rein raised to low hand position–attached to stirrup bars

BACKING

By now the youngster should lunge well on both reins at the walk, trot and canter, and is ready to be backed.

There are many ways of backing a horse—how one does it is entirely a matter of personal preference. I back in a large stable, in the field or in the covered school. First of all, I jump up and down with my hands pulling on the withers; then I get someone to hold my young horse and, while talking to him, give me a leg up so that I lean on his shoulder. I get up and down several times like this, then I am led around hanging on one side to accustom him to the weight. I then get given a leg up and go over his back, patting and talking all the time while leaning with my head down on his neck until I feel him calm and relaxed under me; then slowly I straighten my body because the height of the rider can frighten the horse just as the height of a person frightens the foal. As Robert Orssich so aptly said, 'Why should the horse carry an Eiffel Tower on his back?' Should this towering of the rider prove a problem, then groom your youngster standing on a chair in his stable.

Fig 123.Bouncing off the ground

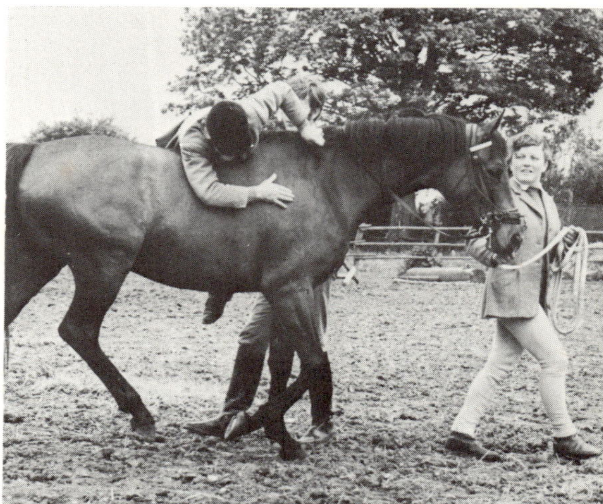

Fig 125. Being led round, weight on the horse's shoulder

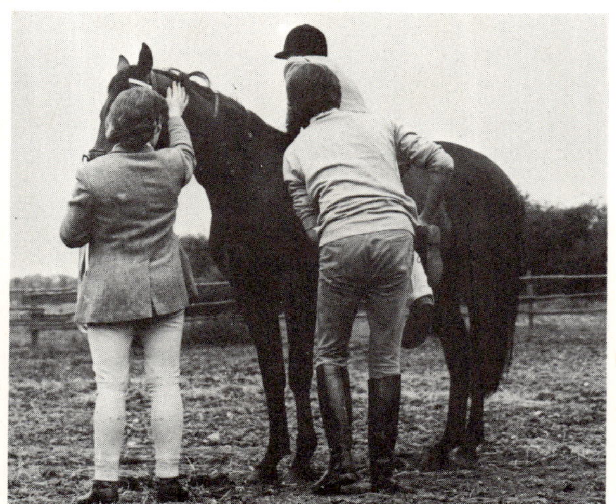

Fig 124. Being given a leg up

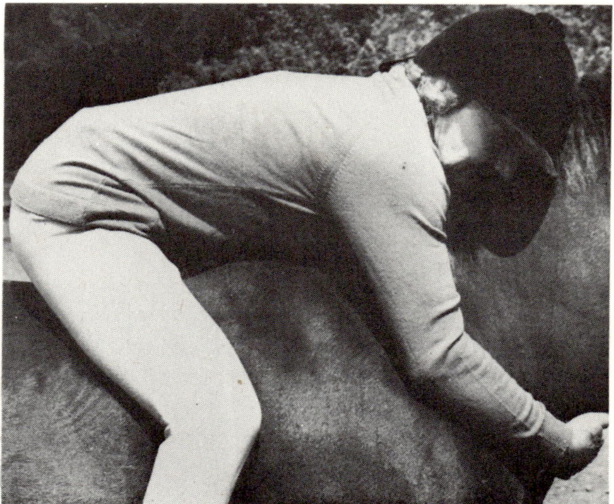

Fig 126. Leaning with head down on horse's back

Fig 127. Riding in a small circle round the trainer on the lunge

Personally I never back with a saddle as I feel the softness and warmth of the human body is more acceptable to the horse's back than the hard cold saddle which could pull or pinch a little with the girth, or at any rate be more uncomfortable, and also the flap could make a noise. Some nervous animals take fright at the slightest things and the less provocation there is the better.

The young horse is now lungeing quietly with a rider astride him and standing still to allow the rider to mount and dismount him. At first the rider has been ballast on the horse's back, getting him used to carrying a weight and being controlled entirely by the person lungeing him in the centre; the next step is for the rider to start to use his reins and legs while on the lunge and gradually to take over control of the horse from the person in the centre of the circle. Soon the horse can be ridden off the lunge in a small circle round the trainer at the end of each work period. This is where the horse who has learnt to long rein finds his work easier than the one who has not, because long reining has taught him to use his mouth and to answer to the reins. The rider must be quiet and confident and transmit to his mount his own calm confidence. The size of the circle should be increased and the pace changed many times until the whole field is used. This is the time to introduce another horse into the work.

Still lunge at the beginning of the work period and progress from one step to the other, always in the same order, so that the youngster will know his work and expect the next stage. The young horse should get used to his quiet companion (never use a companion horse that will give a bad example of behaviour to your inexperienced horse, since horses, like children, seem to copy bad habits quicker than good ones!).

If the trainer has discarded work on the long-reins, then it is a good thing to lead the young horse up the village or down the lanes on the cavesson before taking him out for quiet rides with his sedate companion. If possible, the 'school master' should be stabled next to the youngster you are breaking, as this helps him to gain confidence in the sagacity of his 'school master'. For example: if something frightened the youngster while out on a ride, he would see his 'companion' just walk past without noticing anything wrong, and he would therefore be assured that it was harmless.

Should your green horse be really frightened of anything, get off yourself, touch the object while still holding the end of the reins–sit on it and coax the youngster to investigate, smell and touch the object. Make a point of riding that same way again as soon as possible and keep showing the object to the young horse until he ignores his *bête noire*.

From this point, the young horse should be ridden out first in company and afterwards alone. Continue his schooling quietly, introducing the new things he must learn and always remembering the old French adage, 'calm, forward and straight.'

Fig 128. Riding in a small circle off the lunge

Fig 129. Leading the horse up the village street

LONG REINING

Fig 130. An assistant to control the head

This method of schooling young horses went out of fashion some years ago. Moving with the times, I once gave it up but returned to it again after a while, as I feel it makes a horse use himself and teaches him to use his mouth and answer the aids of the reins. But I would never start by teaching a horse to long rein off the bit for fear of ruining his mouth. I always teach long reining on the nose, which cannot in any way harm the horse's mouth. When a horse has learnt to long rein well off his nose, then he can have the reins attached to the bit. The British Horse Society does not encourage this method of training because, they say, except in the hands of an expert, this method can be dangerous. I agree with this statement, but would qualify it by saying an expert or a thinking and careful trainer. If a horse is taught to long rein from the bit, his mouth must be in danger of being damaged. The other thing that can happen is that the head of the horse can be brought to bend on the circle incorrectly by too tight an outside rein. This must always be guarded against and kept to the forefront of the trainer's mind and eye so that he sees that the correct flexion is maintained on the circles.

To long rein, a cavesson with 3 rings is desirable, although one can manage with a one-ringed noseband by buckling the long reins on to the nose-band in front of the cheek pieces. The horse which is already obedient on the lunge is held on the front ring of the cavesson by the assistant while the trainer takes a pair of lungeing reins or long reins. (The lunge rein has a swivel to stop the rein twisting as the horse goes round the circle and changes direction, whereas the long rein is an ordinary buckle and should have no loop at the end in which to catch the trainer's foot.)

To start with, take one long rein in the hand and with the other rein gradually work it backwards, first on to the neck, then over the back of the saddle and finally round the quarters on both sides of the horse. It is the feeling of the reins round the quarters that usually causes the trouble and one may get quite a bit of kicking from a nervous animal. This is where the assistant controls the horse with the rein on the front ring of the cavesson and talks to and pats the youngster to calm him down until he accepts the rein behind his legs. The trainer can also help by keeping

Fig 131.

Fig 132.

Figs 131-133. The reins passing round the quarters

Fig 134. When using a saddle with stirrups, tie them together with string

his hands wide apart so that the reins do not touch the legs in any way—being careful at the same time to keep himself far enough back not to get kicked. His reins must therefore be long enough to allow for all eventualities. I have two pairs of long reins: a nice light pair, slightly shorter than normal, to use once the horse has learnt to long rein, and a long, strong pair for teaching work on long reins. I think that if you have trouble with a young horse or a spoilt horse there is nothing to beat a lot of walking calmly about, talking to the horse and making him obedient to the voice while you have the reins on the cavesson. Make him obedient on his nose—'Whoa', 'Walk on', 'Get back', etc. Trot and walk, changing up and down on paces. The trainer must be willing to run and walk

Fig 135. A walk through the traffic

and not do only circle work. This is good in its own way, but I feel that work on circles is better done on the lunge where there is no fear of the wrong flexion. It is so easy otherwise to pull the outside reins just a little too much.

I think there is nothing to beat taking the horse up the village on long reins, with one person at his head and one driving on behind him to get him used to what he is likely to find outside the schooling area: to get used to traffic, to face a large lorry with some-one at his head to reassure him with a pat and a kind word. So when he is ridden out with another horse the world is not so astonishing after all. Should the horse persist in dropping his head and fighting the tension of the long reins, then I would advise attaching side reins to the rings of the cavesson (*Note,* not on the bit) and crossing them over his shoulder to keep his head up. When long reining on a quiet horse, side reins can be attached to the bit and back to the dees on the saddle or roller. This helps the young horse to get used to a bit and to mouth himself while at work.

In my youth one often met young horses being long reined down the country lanes during the summer months when we were out riding. I certainly think this method helps any young horse to have a good mouth, in the delicate hands of a good, sympathetic trainer. When he eventually rides his pupil, he will get that lovely response to the indication of the reins in any direction.

RE-SCHOOLING

It so often happens that a horse will not be exactly co-operative after he changes hands. It may be that he has been bought at a sale and wrongly warranted or that he is just unsuitable for the rider who has purchased him. The usual plea is that such and such a horse is too strong and runs away or is being slightly nappy for his new rider, so one is asked to have the horse for a short time to see if one can correct his ways. In nine cases out of ten I start again by mouthing the horse, lungeing him, long reining him and literally starting from the letter A. Some horses respond fantastically to this and, as I write this chapter, there is a certain horse, named *Cosi,* who comes to my mind. Some four years ago he was brought to us by trailer. Neither my daughter nor I nor his late owner could ride one side of him–he seemed impossible to steer or control and I found it difficult to believe that he had ever been hunted. I was asked to school him. He was a bundle of nerves, so I started off by lungeing him for a short time three times each day. I handled him in his box and fed him well–mostly on bran, boiled barley, linseed and nuts and as much hay as he would eat (he was in poor condition). At the end of a fortnight, my daughter was riding him on the end of the lunge for me. By the end of three weeks I was riding him myself, still with fixed reins on him. From there we progressed. He had been jumping on the lunge, but when I took him for a short ride and up to our jumping field the deterioration in his work was most disappointing. So I continued to school him myself, both lungeing and riding, until I had made him a most amenable ride. Then I handed him over to my daughter and his owner, a boy aged 14, and he progressed reasonably well. The autumn came and he really excelled himself in hunter trials, we hunted him lightly and he was a delightful horse in every way. Since then he has been passed on to yet another 14-year-old and has won innumerable classes for working hunters and riding horses and has been placed in several jumping competitions. This was a 10-year-old horse when he first came into my hands, so that I feel most horses will react to re-schooling if enough time and patience can be spared, but it does take longer to re-school a spoilt horse than to train a young one and setbacks are more likely to occur.

Fig 136. The author riding Cosi*–after re-schooling.*

Fig 137. A dropped noseband

To re-mouth a horse I would suggest the introduction of the keyed breaking bit or a treacle-wrapped bit and start back again on the lungeing rein. But it will be found that if you can touch the horse's jawbone with a bit in a different position, so that the place where his jaw is worn and insensitive is not used, he will often be found to have a good mouth. This is why so many of the old school used to use two snaffle bridges, one fitted slightly lower than the other in a horse's mouth. Likewise, a gag or Kimblewick will again touch the horse's mouth in a new place (See fig 201).

Talking of the Kimblewick, I know this is not a popular bit amongst many people but I can only say that we have often found it a most successful bit with ex-racehorses who were far too strong for either my daughter or myself to hold in a plain snaffle.

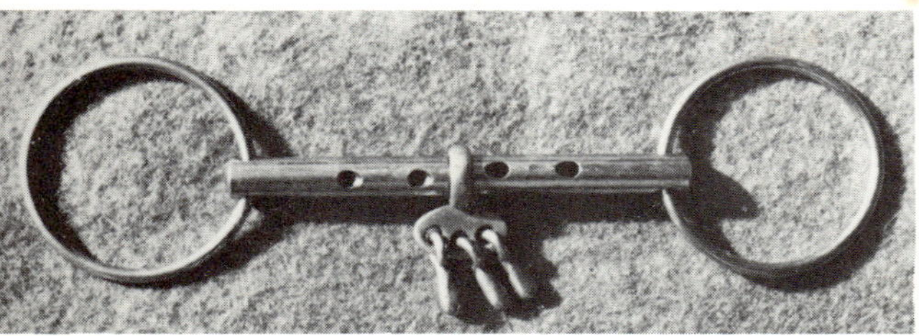

Fig 138. A breaking bit which can be used with or without keys

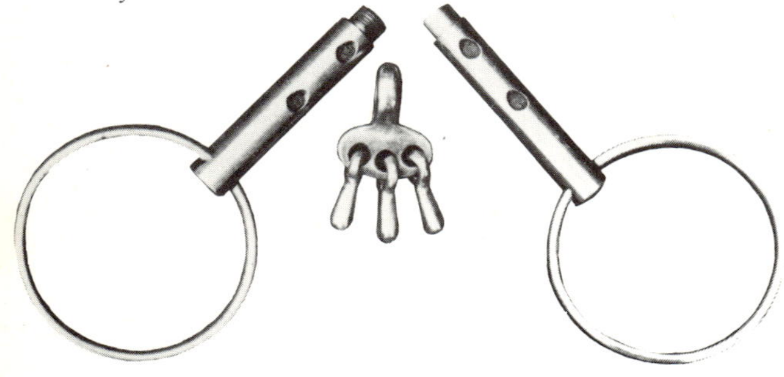

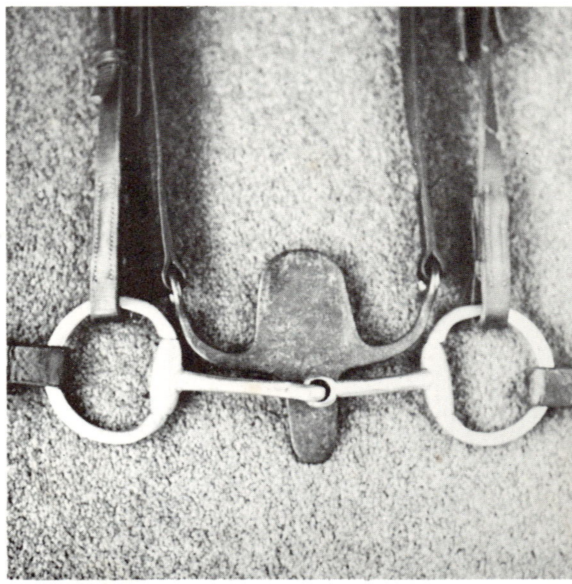

Fig 139. A tongue check–used under the bit

Fig 140. Happy Thought, *owned by Mrs Fane, being schooled for eventing by Miss Clare Newton*

FURTHER TRAINING

I have purposely left the problem of not forwards, i.e. backwards, until there is no problem about understanding the forward movement. Personally, I teach the horse to move backwards fairly early in his schooling, but it must be realised that too much moving backwards can very easily develop into a run backwards when halting, i.e. an evasion. I feel this backwards movement should first be taught on the ground. Place one hand on the horse's nose and one on his shoulder and push him back one or two steps, using the word of command: 'Get back'. Pat him for his obedience and lead him forward again—this leading forward again is most important.

When teaching the rein back with a rider on top, the trainer again pushes the horse back and the rider indicates the aids—weight on his seat with legs slightly closed, flexing of the horse's head by a slight feel on the reins and asking for the rein back with his voice and legs rather than with his hands. When the horse fully understands the rein back, then the rider can do the exercise with the trainer standing by to help if necessary. Some people advocate tapping the horse on each front leg in turn with a stick, or treading on his coronets, but this I feel is unnecessary, inflicting pain or discomfort when it is not needed. Another method is to tap a horse on the chest and use both reins. Again I feel this is not so easy or comfortable as the method of pushing backwards.

The lungeing done with the young horse has

Fig 141. Pushing the horse backwards

Fig 142. Keeping the quarters round on the circle

taught him to work on a circle. Now the rider, when working on a circle, asks with the inside hand for the bend of the head and neck, and by his outside leg slightly behind the girth keeps the horse's quarters round on the circle. (A good way of making sure one rides a correct circle is to get a lunge rein, and sack of sawdust, with someone holding the rein in the centre of the circle while the other person drops sawdust to describe a circle. This is also a useful way of measuring a 10 metre diameter circle for dressage.) Should the animal tend to fall into the circle, the inside leg pushes outwards over the girth and the indirect, or outside, rein eases outwards, not losing the direction of the circle with the inside, or direct, rein. In other words, the rider's inside hand is motionless once the required bend is given. The rider should practise increasing and decreasing the size of the circles he rides. This brings us to better training and asking for more obedience. Likewise, riding on straight lines is very important, and the horse should be taught to keep his body straight by a co-ordination of the rider's leg and hand.

The next movement is sideways as opposed to straight. This can be a common evasion practised by a hot horse going sideways to evade the pressure of the bit. A nappy horse for some reason will run sideways to a gate or towards the stable or other horses.

Before going any further in the horse's training, I would suggest that the rider makes clear to himself the feeling of the correct movement for which he is about to train his horse. He should ride an easy, well-schooled horse at *shoulder-in* on both reins, *pivot* both ways and *half pass* on both reins at the trot and the walk (the trotting half-pass is easier than the walking half pass because the forward impulse is there, while to keep impulsion at the walk plus crossing the legs *forwards* is more difficult). This is where most horses and their riders need help from an experienced teacher who has really done it all. To exercise these movements on the trained horse is easy enough. Just pop your leg in and away he goes—sideways—but with a willing young horse it is another

Fig 143. Teaching lateral aids on the ground

thing altogether. He just does not understand what is being asked and may get confused. To move sideways away from pressure was one of the first things a horse learnt in the stable. This moving over was taught at first with two hands pushing on the shoulder and the middle. As he learnt more, the move over or get over was a mere touch with the hand on the quarters and the horse moved over to be fed, watered or mucked out. So these stable manners now come into the training again.

There are two schools of thought on this sideways movement. An experienced show rider such as Harry Tatlow can make any well-schooled horse trot sideways with no difficulty because of his great natural ability to feel for movement. I remember well one day saying to him that the horse he was riding had never done any sideways movement. His reply was 'But this horse uses both legs at the canter correctly and is obedient to the rider's legs.' Then away he went at a perfect half pass crossing both in front and behind on both reins. Horsemanship for you! But we lesser mortals use more suppling exercises to get the horse moving away or giving to the rider's leg. We practise shoulder-in and -out working on the line of a fence or hedge or the wall of a covered school. Here we ask for the head slightly inwards and legs moving on three tracks, i.e. when the head flexed to the right, the off fore has a track of its own, the off hind leg moves into the track of the near fore while the near hind also has a track of its own. The horse is then moving on three tracks at the rider's orders. In other words, the horse is moving unnaturally at the rider's command. Another exercise for training the move away from the rider's leg and the use of the lateral aids is the *turn on the forehand.* This turn is now out for dressage but is useful when opening gates, while in days gone by it was very important to riders when riding over bridle paths. As a child, I used to ride to see my friends over miles of gated green roads. This is also an excellent exercise for the rider to improve his aids and to use his reins. When it is correctly executed, the horse pivots on one fore leg crossing his hind legs to describe a circle. The horse must not step backwards but one step forward is allowed. This is where we use both lateral aids, i.e. same leg and hand, and diagonal aids–opposite leg and hand. You notice that I put the leg aid before the hand; it is correct that the leg should indicate first and be followed by hand and then followed by firmer leg pressure. This preparing the horse for the movement or exercise is very important when asking for something out of the ordinary. To my mind the lateral aids are corrective aids or forcing aids, i.e. the horse shies away from some object, his head will turn towards the object and the quarters shoot away from the object. The rider, by using the same leg and hand,

Fig 144

Fig 145.

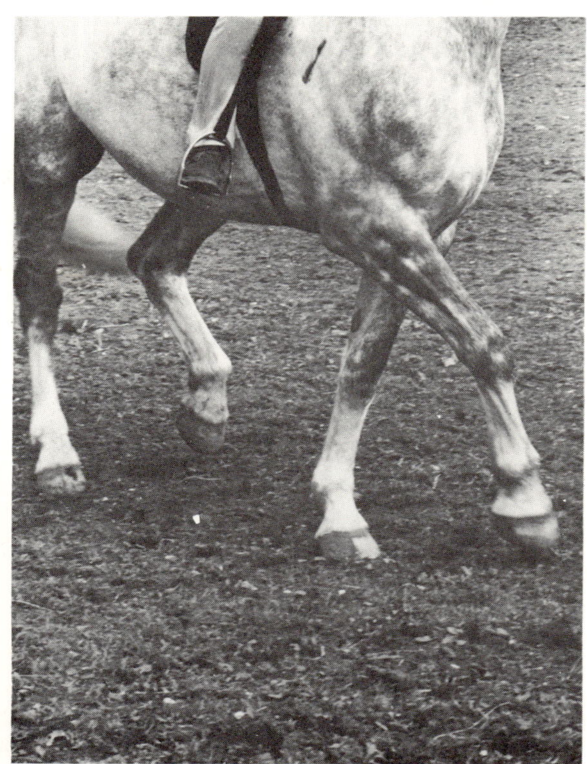

Fig 144-146. Stages in the half pass

straightens the horse so that it can proceed on its way forwards. With these corrective, or lateral aids you, the rider, force the horse straight. You force the horse to move his body in an unnatural way, crossing his legs over. This is where forcing must be gentle and asking more than coercing so that the horse remains calm.

The *half pass*, to use the English term, is a graceful movement when done correctly. The horse moves forward, straight and calm, progressing both sideways and forwards. That is why it is called a two-track movement, i.e. forwards and sideways. When practising more control over your horse you can always teach the *full pass,* that is, to move directly sideways not gaining any forward ground. This is more difficult and is asking a lot of a young horse. Some people who use a confined space for schooling do a lot of sideways work but, I think, mostly out of boredom (although they say it is to supple the horse).

I feel that initial suppling is best done on a circle, working at a walk, trot or canter. Whatever further schooling is done should be done with care and a very complete understanding by the rider since, if badly done, it can cause an immense amount of harm. An over-schooled horse is an unhappy, worried horse, apprehensive of his rider's wishes, fussy in his own brain.

Fig 147. Cantering straight

Fig 148. Horse going well on two tracks

Fig 149. Hind leg into the track of the fore leg

I know one can argue about all kinds of things, but a horse going well straight and tracking up, i.e. hind legs in the tracks of the front legs and correctly bending on his circles, can go ahead anywhere. Any teacher would be glad to have an unconfused, naturally balanced animal. So from here one has to decide what one is going to train for. So often a badly broken or fussed horse comes into one's life, usually by mistake. My advice in such a case would be to return to the lunge. Work on the lunge and most of your troubles will be little ones. Once your horse calms down and goes fully forward then you can start thinking about and planning future training with an ultimate end in view.

SPECIALISED TRAINING

We must now decide for what purpose the horse is to be trained. Often one has some idea what one wants to do with a horse but often one assesses the horse completely wrongly. Once I wanted a jumper for my daughter. We had a long-backed, not very good-looking mare of 15.1½ hands who was hot and explosive. I trained her on the lunge and she jumped well. We won three silver cups in a week at hunter trials and Pony Club jumping. She jumped 45 fences clear on her first day's show jumping. At last we had a show jumper! After three years we found, again quite by chance, that we had a steeplechaser and great was the day she ran away from the field at Fakenham and everyone in the owners' stand was saying, 'What mare is this? Never heard of it!' Unfortunately she did not win as she made a mistake when leading by a fence, but she gamely finished third on three legs. She had a lion heart wrapped up in her little body.

Fig 150. Jennifer Moir (Jinks) schooling Alexander V, *who subsequently qualified twice for the Midland Bank novice events*

THE SHOW JUMPER

The show jumper must prove he is bold and loves jumping. There is no one type to look for, as horses jump in all makes and shapes. But to be able to lift himself he must have a strong back and quarters. A horse may be a fantastic jumper when at liberty and the poor breeder or owner can't keep him in any field. But when broken he can be most disappointing in his ability to jump. This I have proved myself on one occasion. Basic training, and obedience to the rider, is the same requirement as for any other horse. Walking and trotting over poles and small jumps on the lunge and with a rider on top improves a horse's eye and makes him prepared to turn and face his jumps, however small, calmly and confidently. A bigger jump which is higher in the middle encourages a horse to bend his back over the obstacle. At first it is a good thing if one can keep a horse jumping from a trot over many kinds of obstacles starting very low at, say, 2 ft and increasing very slowly as the pupil becomes more confident and takes everything in his stride. It is all-important not to over-face the youngster, but introduce one new obstacle or problem as often as possible. The principle is not to shake the horse's personal confidence in himself. Some people advocate jumping every day, but I think it is sometimes a good thing to do long rides and some schooling on the flat if possible.

Try to jump occasionally away from home, in someone else's field or over the Riding Club jumps in company with other horses in a strange place. Should the young horse become excited there, trot him about and do not jump him, perhaps not at all that day. A day apparently wasted may mean a lot gained in his training. In this kind of training, as in any other kind, try to find someone in whom you have confidence to help you when you are riding over fences. Kindly criticism can be helpful from someone on the ground who can see exactly what you are doing. But it must be someone who you know has an eye for this kind of work. However well the horse is progressing, I beg of you not to put away the lunge rein. It helps so much in balance and calmness.

Fig 151. Trotting over poles on the lunge

This is only a very short piece of advice. Whole books have been devoted to the art of show jumping. Read as much as you can. Every book has a slightly different approach and one may help you more than another. Start with small shows and try for slow clear rounds. Leave the other clever chap to win on time–your time will come. Let your horse enjoy his day's outing. No rosette or prize is worth a spoilt horse. Once he has learnt to rush his fences, and jump with a hollow back, your many hours of slow work will have to be started all over again. Most good jumpers are usually a little impetuous, and that is why slow, calm work pays off. Moreover, the rider must practise at improving the strength of his seat in the saddle. This was shown very clearly when Goonda Butters came to England. She was so firm in her seat, and never lost her balance even when her horse, *Orskiet,* did a most unexpected stop at the Royal Counties.

Fig 152. Rider allowing the bend of the horse

Fig 153. Small jumps on the lunge–up . . .

Fig 154. . . . and over

HACKING FOR PLEASURE

There are many people who have no great ambition to do great things with horses but are content to take gentle exercise on horseback and to enjoy the countryside, probably exercising their dogs at the same time. This is a delightful pastime in good weather. There is no hurry, and the rush and urgency of the present age is unable to hasten the pace. After a long ride in the country we all feel pleasantly tired, at peace with ourselves and the world in general. This peace, which was commonplace in my young days, is a thing to be valued these days. Now there is never time to pass the time of day with the people you see when dashing past in a motor car; the people in the same village to whom one doesn't speak even once a year. The great fraternity of the horse is a thing almost of the past when everyone welcomed the sight of the riders on their horses and we were all on pleasant terms together.

But as one age leaves us another era comes into

being, and of recent years a general interest in riding has come about. The riding holiday has become fashionable, be it in Spain, Italy, Hungary or our own country. Everywhere in the British Isles, children who are horse-crazy persuade their parents to let them spend a holiday in the country with horses or ponies, either by themselves or with their parents. There is provision for almost every taste. There are those who wish to stay in first-class hotels and ride and some who wish to live in camping style in old converted farm buildings and live a simple life with long rides over open country all day and in all weathers. It is amazing how townspeople seem to love this kind of holiday. Other people enjoy their hacking over the week-end at the riding school or evening rides in the summer after work, with a call at a local pub for a drink and a sandwich. Riding clubs organise friendly rides using members' horses so that one can explore the country one lives in or the adjoining counties. This appeals to people who cannot be among the hunting folk and enables them to know their part of the world as seen from the back of a horse. Here we have another aspect of the comradeship of horse owners and those who like riding for riding's sake.

POINT-TO-POINTER

In days gone by the training of race horses or point-to-pointers was not studied very much. A horse was either a good galloper and had a will to gallop or not. He liked either a left- or a right-handed course. But the trainer did not study why he liked one or the other: it was a mere statement. Now it is generally accepted that good basic training to develop natural balance is an asset to every horse, and this will eliminate a lot of changing legs and meeting fences wrongly. So your point-to-pointer should be taught to lunge to improve his balance, to work on his less favoured leg or bad side, and to stretch those weaker muscles of his back and neck. When one considers the breaking of most race horses, which consists of a few days on the lunge followed by riding in a string, the idea of following in a line is instilled into the young horse's mind, and one can understand why so many young horses fail to understand when they are asked to go in front. It is against their basic training which has been to keep behind a leader; this is also true of many good riding school horses and ponies. We have followers (God bless them!) and leaders (Oh heavens, he won't go behind without fussing!). The same applies to a keen horse with a great heart that won't be beaten; he is your real racer with courage to the end.

Fig 155. Before, in not very good condition

Training the point-to-pointer

First, you must find a suitable horse. For the first season, the object should be to have fun and gallop along behind the first flight to learn how to race: to understand the pace and the position in the saddle and that really fast jumping is so different from hunting or show jumping.

Should you have bred the horse you are going to point-to-point, he will have been broken and quietly ridden about in company with other horses (this is where the growth of Riding Clubs has proved so useful to one-horse owners) and hunted quietly as a 4-year-old. Then, as a 5-year-old, when the horse's bones are calcified, or hardened and mature, he can have a few quiet races with the intention of really racing him at 6 or 7 years. It is a slow process to get a point-to-point horse to the right age to do his work. That is why most young riders buy their horses from people who have patience but cannot afford to race. Or of course you can find a suitable horse in a stable yard which is not interested in point-to-pointing and where his capabilities have not been realized.

Now the horse has come to the stables from which he is to point-to-point, sometimes straight from grass or from a sale ring. In most cases he is not in very good condition. He must be wormed, his teeth examined and filed if necessary, and the horse built up to be fat and well in himself. Then he must be hunted eight times with the pack of hounds he is to be qualified with. The secretary will give the owner qualifying cards which are to be presented and signed each time the horse is hunted. Once the horse is qualified he should be half fit, since he has been having two hours' exercise most days when he is not hunting. Training now starts with long slow work on the roads to strengthen and harden the legs; then we can introduce the occasional canter or gallop. The long, slow work can be done at one stretch or in two

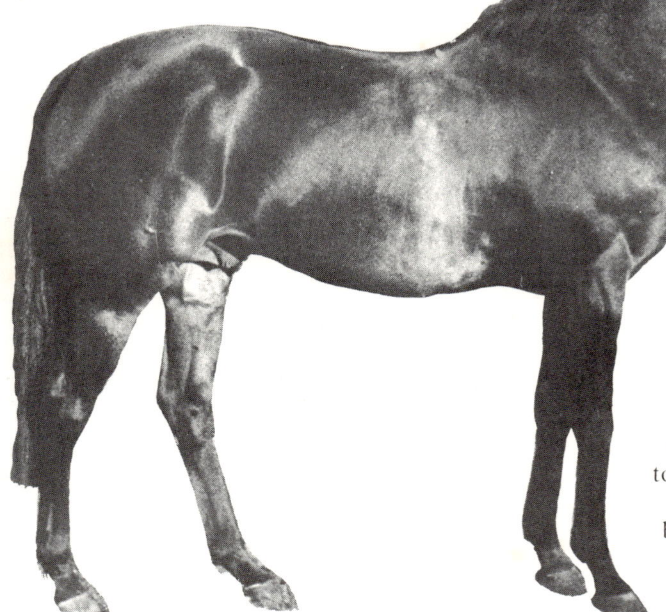

Fig 156. . . . and after good food, exercise and grooming

exercise straight off. There is nothing wrong with this but to give two spells breaks up the horse's long hours in the stable and allows the rider more time to do things in between the periods of exercise, and gives the horse a fresh appetite for his tea. This is when feeds at night should be very late, not before 8.0 p.m. It is nice to join other people for a gallop occasionally as it gives the horse an interest and the owner can get an idea of pace plus the horse's capabilities. The first race will merely be a try-out with no idea of getting to the front. The second race will be a try for a place. It is after this that you must decide whether or not the horse is able to make the trip, in other words get the distance. Perhaps he is not able to keep the pace or the distance. You have backed the wrong horse and another sphere in life must be found for him, perhaps as a hunter, an eventer or show jumper. The programme for the last four weeks of a point-to-pointer's days should be roughly on these lines. He should have three hours exercise a day and about every two weeks a canter of three miles at half speed. One or two pipe openers, that is a short fast gallop of one mile or so, and one fairly fast three miles about four days before the race, followed the next day by one hour's walking. A rest day before the race, walked out in hand to pick a little grass if possible. Then he should be fit and hard, and well in himself to give of his best. The great trouble is to get 16 to 20 lb. oats down him daily, helped with some bran and grated carrots. A few eggs and stout or black beer may be added to his last feed during his last week before racing.

After racing the horse needs a good salt bran mash to replace the natural salt lost in sweating. The next day a walk out in hand and a laxative diet if possible as a change. It is nice to win but it is nicer to have fun and enjoy your horses and experience the comradeship of those sporting people who love to see good horses galloping and jumping and to watch the younger generation grow up enjoying the thrill of racing. The thud of hooves never fails to quicken my blood, having been introduced to racing by my father who was, in his day, a well-known point-to-point rider. But I have learnt that there is no short cut to getting a horse fit and that slow work pays dividends.

sessions: usually the latter suits most horses and riders. Give two hours early in the morning. After a short feed, you must wait at least one hour before leaving the stable and this is where the old hours of stable work (6.0 a.m. first feed) pay off. You can go out as soon as it is light, or sooner if you can get off the roads quickly and the horse is sensible. I have had some lovely early morning rides later in the season watching the dawn come over the hills and the sun rise, when the birds are just waking and people still asleep. The countryside comes awake round you and the world is so peaceful, just the trot-trot and sound of the horse's feet on the wet earth. Sometimes, it is so cold that the horse's breath is like smoke and you have no hands or toes at all.

On return the horse is given a few minutes to drink, roll and stale, pick a little hay and settle down. Then give him a good feed helped with, for example, grated carrots or molassin meal if the horse has a sweet tooth, or a little salt if he prefers it. Then a good strapping and wisping which, they say, is as good as a good feed. Rest, followed by a light mid-day feed, then rest for one or two hours and another hour's exercise. The old idea was three hours'

SHOWING

Should one be interested in showing and have been lucky enough to have bred or found a good type of animal, the basic training will be the same as any other horse, i.e. lungeing and riding out in company to make him a quiet, obedient ride. But now we have to think more about carriage and posture. This horse must be able to attract the judge's attention. Now just how can this be obtained? Having been both judge and exhibitor of show hacks, hunters and children's ponies in-hand and under saddle, I think that the way a horse carries himself is the first thing that makes me notice him. He must be, as Robert Orssich said one day, 'Like a beautiful woman who enters a restaurant, head held high and poised. As she walks to her table, everyone puts down their knives and forks and stops eating!' Nobody would have looked up had that lady come in with her head held low and poking forward. So, foremost, the horse must have an elegant and gay head carriage: the neck muscles must be well developed and the head held high and flexed from the poll, no chin sticking forward. As all horses in hand or ridden are first seen at the walk, this gait must be well measured, with a good stride. This pace is important especially with a hunter. When one thinks of hacking to a meet or walking home or back to the box after a day's hunting, nobody wants to have to push a horse to make him walk when both the rider and the horse are tired. Nor is it nice to have to push a horse at the walk or if you are hacking in company.

The best way to develop the walk is to lead the horse in-hand from both the near and off side to develop his muscles equally on both sides. Walk out really well yourself and push him on with a flick of a long whip carried backwards in your hand. Be sure that the bit is correctly held, i.e. under the chin, if he is led in a bridle (although I prefer a three-ringed cavesson myself). When practising walking with the rider on top the legs should work alternately with the stride of the horse—first on a loose rein and afterwards raising the head higher until one gets a naturally

Fig 157. A good natural head carriage

good carriage without shortening the stride or length of reach of the foot.

Trotting down a field with a slight decline always helps the trot for ridden work. But really the trot can best be improved on the lunge, with the rider being careful that the hocks do not trail behind as the horse moves forward. Truly good movement at the trot can only be attained through the hocks coming well under the body to take the weight and allow the forward movement of the front legs. For the canter, again work on the lunge will give the effortless calm canter which one wants to see. In other words, as work progresses and the head comes up, we get more natural collection or poise—that is, the shortening of the horse. When working with young horses, ridden or on the lunge, one catches glimpses of what one hopes to attain after months of work. This gives the trainer the heart to go on trying and working. There is an old saying, 'You can judge any horse at the walk', which means that a good walk on a well-made horse indicates in 99 cases out of 100 that that horse can gallop.

By gallop I do not mean to scuttle round the ring the way that 75% of the present-day show ponies do in their individual shows (more the pity that the gallop was made part of their show!) but to stride out and open out, lowering the body towards the ground as they go forwards. Some horses take time to develop this movement in a confined space, or a badly-shaped ring. These horses need more balance and more practice in going into the gallop and back into the canter and raising and decreasing their pace from canter to gallop and gallop to canter without fussing.

There are, of course, those horses whose best pace is the gallop. There are also those horses who need a clever rider to get the best out of them; a rider who knows the tricks of the game and can understand ringcraft, not riding dirtily but riding cleverly, showing his horse off to the judge at its best, making the most of his good points and minimising his bad ones. One professional rider was waiting at the entrance to the ring one day when someone said, 'Go on into the ring.' 'No,' he replied. 'I'm waiting for a bad one to follow!' There is no denying that some people make a horse look much nicer when they get on top. I always said that Larry Mooney improved any horse, however common. Its head used to come up to his sympathetic hands and it was worth £50 more just by having Larry on top!

Fig 158. The lunge will give the effortless calm canter

No horse is perfect in conformation and if you wish to show him you must consider his faults and try to minimise them.

Should the head be rather large or coarse, do not put on a small nose band. A wide one will improve the look of the nose and the length from eye to nostril, while a narrow nose band will accentuate the commonness. If the ears are on the large side, then trim them carefully as the long hair in the ears makes them look larger. If the neck is coarse, then sweat the extra fat off it by bandaging a piece of rubber (an old tractor inner tube is very useful) on to the neck whilst working. Should the neck be too thin, then the judicial use of side reins can improve the line of the crest of the neck by making the horse flex and develop his muscles. The shoulder cannot really be helped except by using a straight-cut show saddle which gives the optical illusion of the horse's having a better front because the rider is sitting further back in the saddle. The middle piece should be studied. Some horses need to be very fat to look well; others need to reduce their waist line. The quarters must be round with no poverty marks on them; these are the lines on the back of the quarters, the bigger the quarters the better. This can be helped by wisping the quarters to bring the muscles up. The bone structure of the legs can be helped by trimming if the horse has a coarse bone and by keeping as much hair as possible on the cannon bone should the horse be lacking in bone. All heels must be trimmed except in the case of Mountain and Moorland breeds, where the feather is left on. But even in registered ponies the hair should be tidied carefully. The aim must be to make the best of what you have because you have to make the judge like your horse well enough to pull it in higher up the line than it should be.

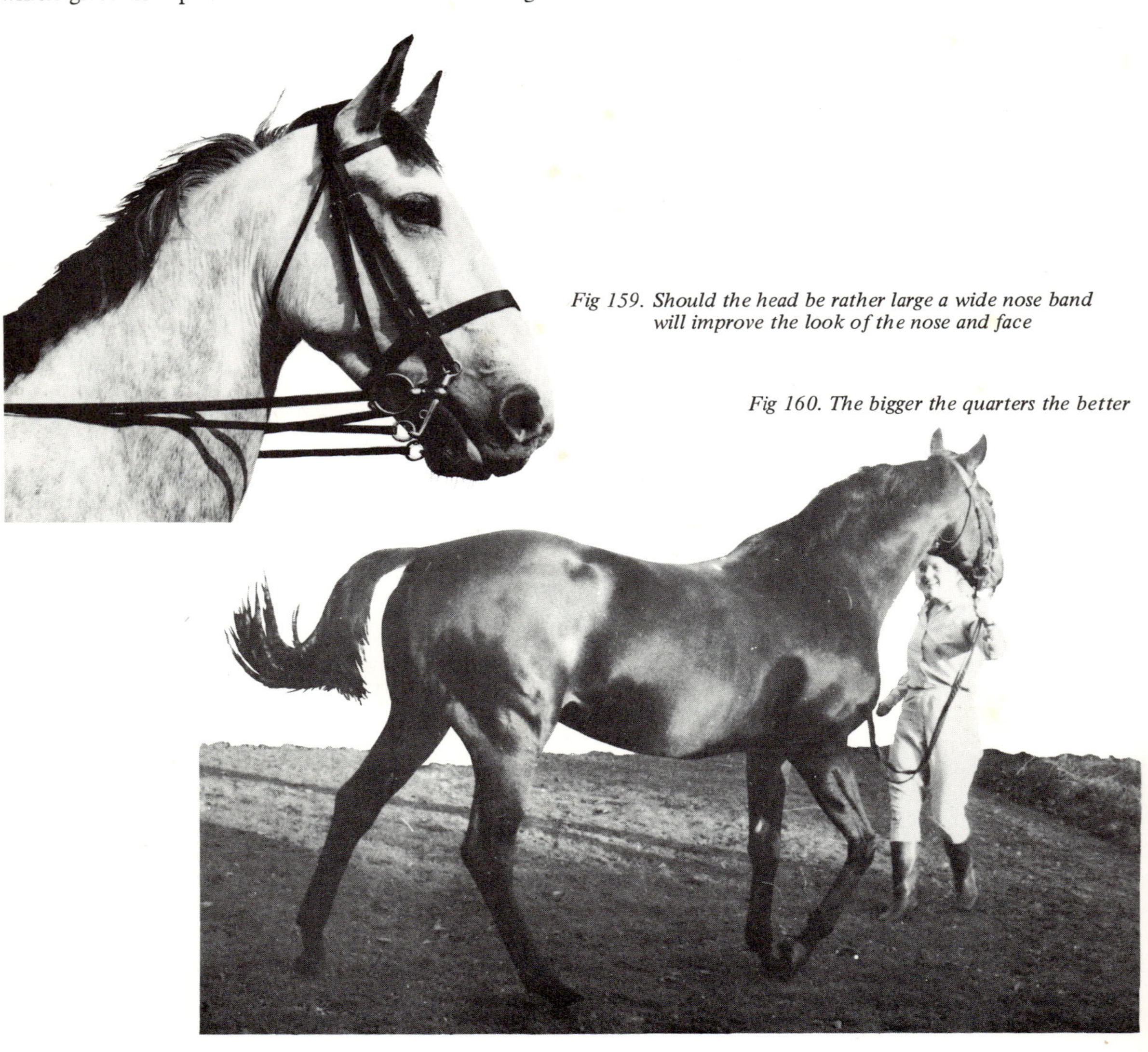

Fig 159. Should the head be rather large a wide nose band will improve the look of the nose and face

Fig 160. The bigger the quarters the better

THE HUNTER

Fig 161. The type of horse that is suitable to ride to hounds– Scarlet Gaunt–*bred, produced and ridden by the author*

A hunter is a type of horse which is suitable to ride to hounds. He may be able to carry over 15 stone, or he may be a small horse, suitable for a teen-age girl. There are now so many hunter classes that every type of hunter is catered for, even in the young stock classes now. A show hunter is meant to be a model of a horse and need never have hunted, in fact some winners of hunter classes have never seen hounds or a covert side. I, for one, do not blame the owners for not hunting a valuable show horse who could halve his value by bumping his knee badly on a stiff fence. One can get no joy of carefree abandon when one has to be careful of where one's horse puts his feet all day and wrap him in cotton wool. Nobody except a millionaire can honestly say £1,000 does not matter, and now there are plenty of working hunter classes for those honest working horses who really hunt and collect a few honourable hunting bumps and lumps. So why bother to run the risk of damaging a show horse? Heavens, they are difficult enough to find.

Usually, at the bigger shows there are four classes and sometimes more if novice and ladies classes are added. The lightweight class is for horses able to carry up to 12 stone 7 lbs; the middleweight class for those to carry between 12½ and 14½ stone, and the heavyweight for those who can carry over 14½ stone. The class for small hunters is for horses under 15.2 hands in height. The horse must be workmanlike, sensible and look as if he could be hunted and be able to gallop on to cover the ground and give the judge a good ride. So he must have a good mouth and not pull and yet go freely forward for the judge, i.e. for a strange rider. This is where the show horse must be able to carry different weights of rider, so his training must include many people riding him so that he does not become one-rider-minded. The working hunter will be asked to jump a few natural-looking fences but the trainer should bear in mind that many courses do include white gates and coloured walls. At Windsor one year a viaduct wall

Fig 162. Junior class ponies–12.2 hands and under

was included in the course. This would hardly be met out hunting! So it is a good thing to school the horse over show jumps and to be prepared for any eventuality.

While talking of working hunters, one could include the now very popular classes for working ponies ridden by young people under 18 years of age. Ponies in the senior class must be under 15 hands; in the intermediate class they must be under 14 hands and ridden by riders of 16 years or under; and in the junior class they must be under 13 hands and ridden by riders of 14 years or under. The ponies have to be of hunter type, i.e. with enough bone and substance to be able to carry out the duties of a real hunter and to be able to jump as well. Their jumping ability is tested on the day by doing a small course of supposedly natural fences. The judge does not ride these working ponies but I feel this aspect of the test could change and at least the bigger two classes could be ridden by judges of 8½ stone and under. But of course that might cause trouble about ladies' weights! The smaller class should never be ridden by an adult, as some of the best small ponies I have known would not carry an adult at all while nursing a child along most carefully.

The next type of show animal could be the child's pony. This is a type on its own again. One is looking for a beautiful model of what can be called a child's mount. These ponies used just to happen 20 years ago. There were a few wonderful ponies and a lot of very mediocre ponies. Now, the breeding of children's show ponies has become an enormous trade, for export as well as for the home market. There are many big studs breeding only ponies and there are a great many lovely ponies at shows. So now there are few ordinary ponies but many little models of different heights. They all have to be under 14.2 hands. The senior class is for ponies under 14.2 and over 13.2 ridden by riders of 16 years and under. The middle class is for ponies between 12.2 and 13.2 and the junior class for those under 12.2. In this class there are two sub-classes, one for ponies under 12 hands to be ridden off the leading rein, called a first pony class for riders under 10 years, and the leading rein class for ponies under 12 hands and riders under 8 years old. So the young are well catered for and any animal under 15 hands should be able to find a class suitable to its height and type.

Ponies of native breeds also have special classes in the show ring. So those wishing to show their

Fig 163. This pony is happy for his young rider to play with him

horses, or, shall we say, go to a party with their horses, can easily have lots of fun, tidying and grooming their horses to make the best of them, besides taking pleasure in buying smart rugs, bridles, head collars and ropes. The demand for saddlery has increased enormously recently. The number of firms selling saddlery amazes me—rather like the number of pubs in some towns! How they can all find enough customers is a wonder. But they must do, otherwise they would not stay in business!

In the show world there are also classes for young animals grouped by ages—1-, 2- or 3-year-olds, i.e. too young really to work or carry a rider. This applies also to driving horses, e.g. the Hackney Horse Society which runs a breed show every year. These young animals are shown in hand and are judged on conformation and the judge has to decide how he thinks the animal is likely to develop.

Now we turn to a not-so-beautiful pony but, to my mind, a more useful pony—the kind that wise parents will buy for their children to enjoy. I bought Prince for my grandchildren. He is not good-looking but he's got good sense. So much so, that he is a marvellous instructor, because he is wise and has great experience. He jumps wonderfully but never hurries, even out hunting. He gives everyone great confidence and pleasure and reassures those who have been a little frightened.

The training of the ordinary child's pony is the same as for any horse except that, after the animal has become confident in its ordinary work on and off the lunge, a really lightweight rider must continue its schooling. It is up to this lightweight rider to teach the pony to be a child's companion and to encourage the pony to take everything in its stride; to teach him not to be afraid when a young rider plays with him as a toy (sometimes not very gently) or practises some feat the owner has seen on television or read of in some horsey story, or slides down the tail. A nervous pony is never suitable for a child to play with. This training takes at least 6 months and makes the pony almost as valuable as the good-looking pony, to my mind. But it's a vain life, and people pay for good looks and want to be better than the other chap. Only at hunter trials do we see the real children's ponies taking their riders over fences—and sometimes, heavens, haven't the organisers thought up some fences?! They would frighten the very foxes who are supposed to cross them! I feel all foxes and most of us of the past generation would go round them!

THE HACK

The perfect hack or horse was never born and, as my father used to say, it would take four good horses to make a perfect one. First of all a hack must be an elegant, eye-catching animal in every way. He must be between 14.2 and 15.3 hands. Sometimes there are two classes for hacks, that is, small–14.2 hands to 15 hands and open. Other shows have only one class. Presence is very important along with good movement, conformation and bone, which of late has been somewhat lacking. When the height was increased to 15.3 hands from 15.2 hands, more of the riding horse or hunter crept in and the truly elegant, strong-backed hack seems not to be found these days. The tendency now is to have the more shelly type Thoroughbred in the ring, for if the horse has a bit of substance or strength about him he is said to be 'of hunter type'. Should he be just over 14.2 hands, with a fine head, he is termed as a pony in a hack class–so there is no pleasing everyone!

In days gone by there were some fine animals of true hack type. *Ebonita* and *Vanity Fair* were two very elegant mares of true hack type. *Liberty Light* was a large hack of Thoroughbred type who had poise and movement. A hack is most commonly a cross-bred animal who 'just happened'. You can call it fluke breeding or what you like, but Miss de Beaumont has been very successful with the breeding of hacks from *'June'* and her descendants. To name a few: there are *Jupiter, Ladybird,* and the latter's daughter, *Last Waltz.*

The training of a hack must concentrate on producing a balanced, pleasant ride–a horse who is willing to go for all judges as well as for the rider he knows: who takes things in his stride. He must therefore be ridden out in order to come across different people and different situations as well as being schooled to make the best use of his paces. This should first be developed on the lunge and further developed with the rider on the lunge. After this he should be ridden off the lunge with side reins still on his bridle, slightly looser than the rider's reins. This steadies the horse's head. I know this sounds nonsense, but I had read it in an article in *Horse and Hound* by the late Princess Frederick Carl of Prussia, and since I had bred from her horse *Don Juan* I was doubly interested. So I tried it out in my own covered school and, to my amazement, it worked. I have since used this method a great deal. The side reins must be unclipped at intervals and the horse allowed to stretch his neck downwards and forwards.

Even when the horse is doing well in his classes he should still be lunged as well as ridden, as this will develop poise. When I was producing the champion ponies *Chocolate Box* and *Picture Play,* they were ridden and lunged on alternate days to keep their schooling going, thus producing better balance and more effortless elegance. This avoided the horses being bored to tears with continual going round in circles.

Fig 164. Last Waltz, *one of Miss de Beaumont's most successful hacks, broken and produced by the author and her daughter, Jinks. Subsequently shown by David Tatlow.*

Fig 165. Chocolate Box, *one of Jinks' most successful ponies, and later a brood mare at Harroway House*

Fig 166. Princess Frederick Carl of Prussia on Don Juan

SHOWING IN-HAND

This type of showing has increased enormously during the last 20 years. Before, hunters and cart-horses were shown in-hand, but as interest in the breeding of all types of horses has increased so has the demand been met by most shows for classes for showing young stock of different sizes. Now there are few, if any, horses who cannot claim to have special classes for their own particular breed. Brood mares are usually shown with their foals at foot, and a very delightful class this makes, with some grand old matrons and most charming small foals. In this case both mare and foal have to be well turned out, with plaited manes and tails, so a certain amount of practice or training has to be done at home so that they will run up well in-hand for the judges at the show. The next classes are for 1-, 2- or 3-year-olds: these can be lumped together in one class with a special prize for the best yearling, or, if sufficient entries are likely to be received, each age group has its own class. Here again your horses must be kept in, well rugged and schooled to run out in hand, and stand correctly. But one often finds that, however much training has been done at home, the excitement of the show in a strange place amongst many horses is too much for most youngsters and they do not remain calm until they have visited several shows and become a little more accustomed to the exciting surroundings. In all the above classes, the horses are judged on conformation and type, the latter being very important in breed shows. Here the judging is merely the personal opinion of the judge concerned and it is most interesting to watch different people judging the same animals. Should you be exhibiting and a great many judges like your animal then you can be fairly certain it is a good one. On the other hand, a champion from the week before can very well fail to attract a judge a week later. The whole object of these classes is to encourage breeding from animals of good conformation.

Fig 167. Breed from animals with good conformation– Friday's Child *has bred seven winners*

HUNTING

To school a young horse for hunting needs patience and time. Usually the horse will be about 4 years old or just over. The ideal is to ring up the Master of your local hunt and ask him if you may bring your 4-year-old out cubbing. No Master really objects to this as long as you keep a young horse well away from his hounds so that no hound will get kicked. This is also to your own advantage, then the horse can look on and not really partake in any real activity. So, having arrived at the meet slightly early, position the young horse well away from the hounds and field but where he is able to see what is going on. If it is possible, it is always a good idea to take some sedate companion with you so that the two horses can stand quietly together and watch the hounds and horses arriving. Then, following at a discreet distance, watch the hounds draw the first cover. Should the young horse show any signs of fidgeting, walk him up and down rather than try to make him stand still. Should he become excited, then walk him away some distance, turn round and walk him back again. After a few mornings spent at the walk and a gentle trot, a short canter can be indulged in as long as the horse is kept well behind everybody else. If you wish to produce a well-mannered hunter it is wise never to allow the horse to take part in real hunting until after Christmas, so that he has thoroughly learnt that his place is to stay behind and not to be headstrong and just gallop away at the mere sound of the hounds giving tongue. In this type of work the old quotation 'Manners maketh man' is very true and many horses are spoilt by rushing them into hunting and galloping off with the field before their discipline is complete. A horse that kicks either hounds or other horses is a disgrace to the person who has schooled it.

Fig 168. All ages can enjoy a day's hunting

NATIVE BREEDS

Fig 169. Highland pony Glengarry

Fig 170. Shetland pony Wells Esta

Fig 171. The Shetland pony wore boots to prevent him damaging the grass

While talking about riding in the British Isles, it would be quite wrong not to mention the nine native breeds that we are lucky enough to inherit. They are the indigenous breeds which have run wild on the mountains and moorland for many centuries. Fresh blood has been introduced to improve the stock from time to time and to counteract in-breeding. It has also worked the other way—the introduction of native pony blood has done so much to improve the stamina of riding horses in England. Many rather small hunters, Pony Club ponies and even High and Mighty (winner of Badminton) have a lot of native pony blood in their veins. More recently the popularity of trekking holidays has created a new demand for the larger breeds such as the Dales, Fells, Highlands and Welsh Cobs. These heavier type ponies were decreasing at an alarming rate because there was no demand for them; the small hill farmer went in for tractors and Land Rovers rather than use these ponies which could plough as well as take produce to market.

I will outline each breed very briefly, giving its habitat, characteristics and general use.

Shetland ponies are the smallest breed and come from the furthest North. Being very powerfully built, they are like miniature cart horses in stature and could carry loads of peat back to the croft and pull small carts. Of late, however, there has been a trend to breed a lighter type of pony and some most attractive miniatures have become very fashionable. Because of their size they were used as children's leading rein ponies and often driven by ladies in the last century, besides pulling the old-fashioned lawn mower. This breed has proved very popular on the

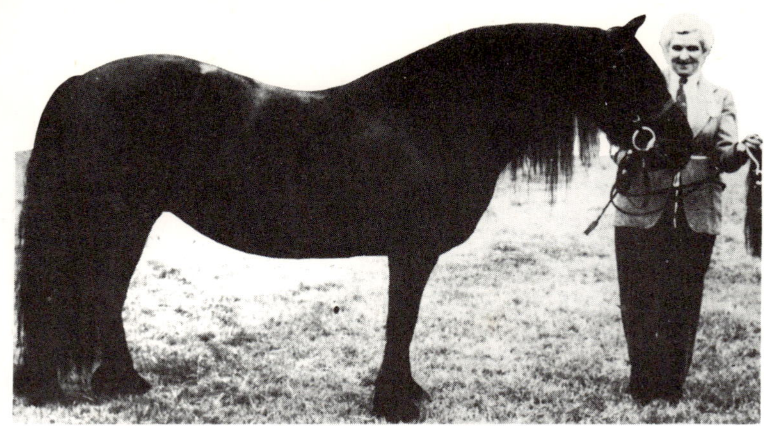

Fig 172. Fell pony Dene Fire Fame

Fig 173. Dales pony Wharton Beauty

Continent and in America. The most favoured colour for showing is black but a lot of the smaller ponies are chestnut, grey or skewbald. Since they are so small they are difficult to school properly as riding ponies, but as driving ponies they are wonderful.

The *Highland* ponies can be divided into two types. Those from the Western Isles came originally from the Isle of Rhum; these ponies are smaller and lighter in type than the Garrons, which come from the Central Highlands of Scotland, are heavier and so well suited to working on the hills deer-stalking and, in the past, ploughing and doing general farm work on the crofts. They can be of any colour other than skewbald or piebald—a lot of them are duns with eelmarks down their backs and zebra markings on their legs.

Fell ponies are usually between 13.00 hands and 14.00 hands. There is a very definite breed type: no white markings are allowed, they are usually blacks, some bays and very rarely a grey, with long manes and tails. They are very game and active ponies and usually a judge in Mountain and Moorland classes is sure to have a good ride on these amiable-tempered animals.

Dales ponies were originally bred to carry lead from the Pennines to the ports and to work on the hill farms. This breed was formed by crossing the Fell ponies with a heavier type of cart stallion, usually a Clydesdale. The colour is mostly black, some bays, two white socks are allowed and a star on the head. These ponies were bred to walk well with heavy loads: they are active and generally very kind and sensible, and are now used extensively in some trekking centres. It is very impressive to see, as one does from a centre in the Lake District, a string of ten or twelve of these animals leaving on a trek together.

Fig 174. Exmoor mare Redwing

Fig 175. A Dartmoor stallion Hisley Woodcock

The *Exmoor* is probably the most pure of any of the native breeds; the ponies are of very definite type being under 12.2 hands, browns or bays, long flowing manes and tails, small ears and the well-known mealy nose. They are really tough ponies, well known for their strength, who have found a living on the moorlands of Exmoor for generations and can often be seen carrying farmers, shepherding and hunting. The first cross with a Thoroughbred makes an excellent, tough, boy's hunter–I say boy's hunter as they are inclined to be wilful but can go all day long and never seem to tire.

The *Dartmoor* is a charming small type pony with the usual long mane and tail, a compact, strong, short-legged animal up to a lot of weight for its size. The usual colours are bays and browns, with some chestnuts and greys. No coloured pony is eligible for registration. Over the years some Shetland and Welsh blood has been introduced and there are still signs on Dartmoor of Shetland blood when one sees skewbalds running wild. They were bred for the coal mining trade some fifty years ago. The Dartmoor cross Thoroughbred often makes show ponies having both quality and delightful pony heads.

The *Connemara* pony comes from the West coast of Ireland and was once a useful type of 13.2 hands pony but with the introduction of Thoroughbred and Arab blood it has become larger and shows more quality than it did twenty years ago. These ponies are suitable in weight to carry both adult and child and are usually good hunters and jumpers. Any colour except skewbald and piebald is acceptable for registration: there are many ponies who are Palomino or cream. The Irish Connemara Society accept ponies into their Stud Book on inspection of type, and there is also an English

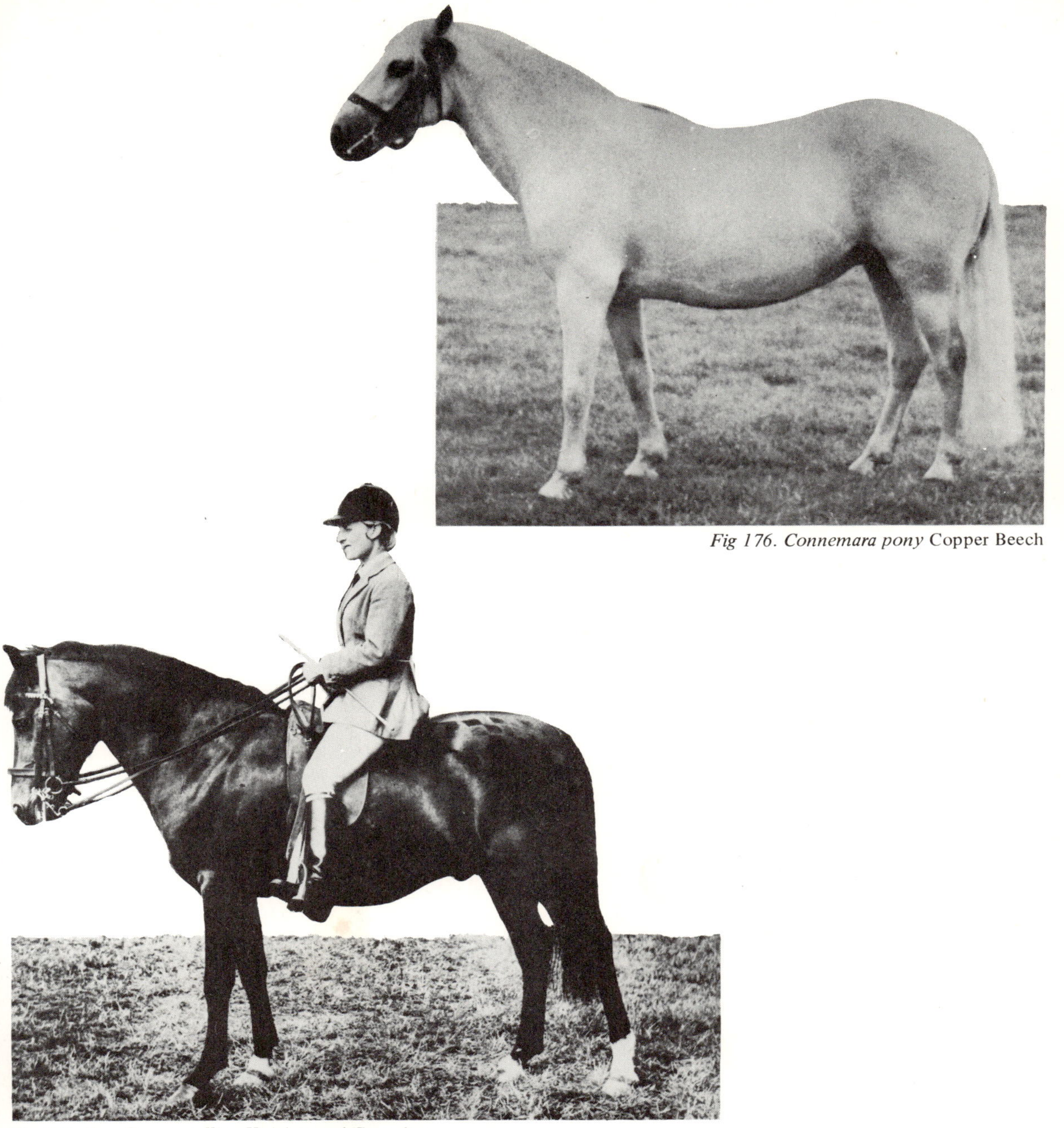

Fig 176. Connemara pony Copper Beech

Fig 177. New Forest stallion Knightwood Crusader

Connemara Society with its own Stud Book.

New Forest ponies come from an area in Hampshire running roughly from Southampton to Bournemouth on the coast, and some 15 miles inland, and graze the rough common land which is a mixture of moorland and woods. They were strong ponies about 13.2 hands, suitable to work on the commoners' small farms and to be ridden. Nowadays there are many studs away from the New Forest where registered ponies are bred as there has been a great demand for export to Holland and other continental countries–with a good temperament and showing ability to jump, they make ideal children's hunters. At the Breed Show there are classes for ponies 13.1 hands and under, and over 13.1 hands and under 14.2 hands, and also classes for New Forest crosses, i.e. half New Forest and half some other breed, in most cases either Thoroughbred or Arab blood.

Fig 178. Welsh Section A Mare—Coed Coch Siaradus

Fig 179. Painting of Welsh Pony, Section A Stallion, Coed Coch Glyndwr

In Wales, there are several types of registered ponies and a stud book is kept for the four sections. *Section A, Welsh Mountain Pony*, is one of the most attractive of the Mountain and Moorland breeds with a pretty head and small ears, yet it should have good bone under the knee. Their height is 12 hands and under, they make excellent children's ponies but, being full of courage and quality, they are inclined to need careful handling when being broken. *Section B, Welsh Pony of riding type, 13.2 hands and under* should show the same characteristics as the smaller type of pony, only being bigger it can carry a larger child and have that much more scope. It makes a first-class riding pony, can jump and hunt to which nothing comes amiss. *Section C is a heavier pony of Cob type under 13.2 hands.* There has been an enormous increase in the breeding of this section in recent years as it is in great demand at trekking centres and has become very popular in the show-ring. It is well suited to carrying light adults and children. *Section D, Welsh Cobs* should have quality heads and small ears, stand up to 15 hands, and have very powerful bodies and great action when trotting. Their manes, tails and feathers should be of fine-textured hair. They are up to untold weight yet, being quiet and amiable, they are suitable for children as well as adults—in fact they are a real family pony.

The Welsh Pony and Cob Society also has two registers. The first is for pure-bred Welsh geldings which can be in any of the four sections and therefore can be shown in Mountain and Moorland classes. The second register is for part-bred Welsh ponies of registered descent; some shows have special classes for these very attractive and useful ponies.

Fig 180. Welsh pony, Section B. Brockwell Cobweb

Fig 181. Welsh Cob, Section C. Turkdean Cerdin

Fig 182. Welsh Cob, Section D. Llanarth Brummel

BITS AND BITTING

When talking about bitting, the most important consideration is that the horse must be comfortable in his mouth.

There is a very true saying that there is a key to every horse's mouth, and equally true that the trouble is to find it. Another accepted axiom is that usually a milder bit is the right one. I think there was a time when horses were ridden and driven in very long curbs, as can be seen in so many old prints, but now there seems to be a complete swing in the opposite direction and everything has to go in a snaffle. There also seems to be a 'thing' about rubber bits and bitless bridles. I feel there must be a happy medium, bearing in mind that the sensitive membrane of the jaw must be guarded because once it is damaged it can never be totally repaired. When the sensitivity of the mouth is lost, the horse is said to have a hard or bad mouth.

When choosing a bit for a horse one must consider the size or width of the mouth and also the thickness of the tongue. One horse will go well in a certain bit for one rider but not so well in the same bit for another because their hands differ, so that one finds different bits suit the same horse according to the rider. The late Major Faudel-Phillips used to keep a great variety of double bridles and his school horses used a different one every day which, to his way of thinking, kept the horses' mouths more lively. I certainly think a change is good at times but also feel strongly that if a horse goes well in a bit then there is no reason to change it. For example, should the young horse go well and like a half-moon or straight bar snaffle, then I see no reason to put him in a more generally used jointed snaffle, although this would, to some people, mean progress in his training.

In the days of the Institute of the Horse (which was the Society recognised by the horse world in England before the British Horse Society existed) there was a very good general question in their examination about bitting which I feel was instructive. Usually a pile of bits was put out on a table and the candidate was told to put them into the order of severity. The rules roughly were rubber before vulcanite, the thicker the steel bit the milder, the half-moon was kinder than the straight bar, the smooth bit was softer than the serrated, the jointed bit gave nut-cracker action, the double-jointed bit had more action on the jaws than the single-jointed bit. Rollers going round the bit had more action on the tongue than those going across the tongue. When curbs and Pelhams were taken into consideration, the same rules held good as for the snaffle but there was the length of the cheekpiece and the height of the port to be considered. The length of the cheekpiece

Fig 183. Cheeks of the snaffle connected to the cheek piece of bridle

Fig 184. A chain snaffle

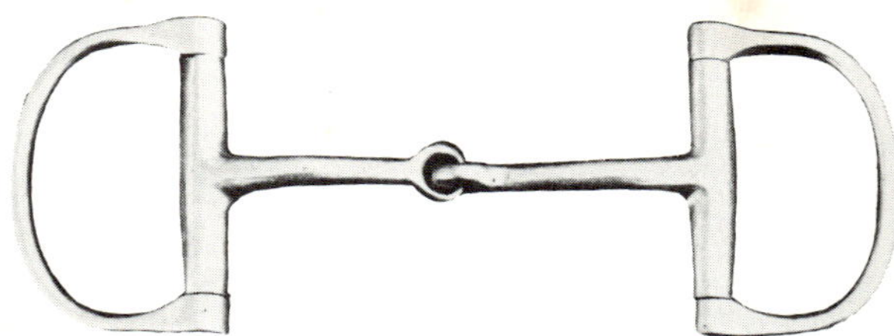

Fig 185. A Racing D snaffle

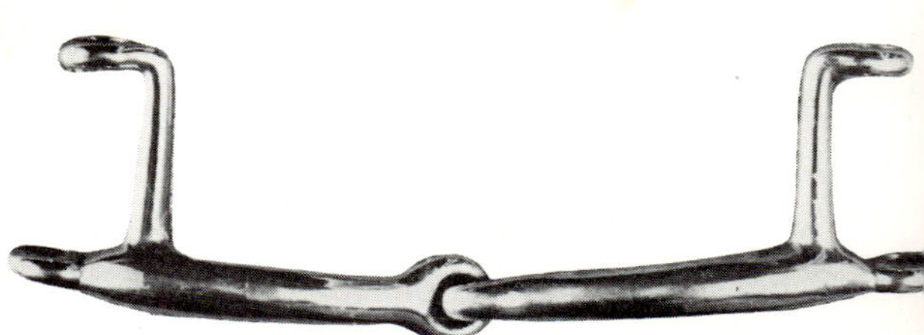

Fig 186. A Bridoon gag snaffle

Fig 187. Half moon rubber snaffle

Fig 188. An Egg butt snaffle

Fig 189. A plain ring snaffle

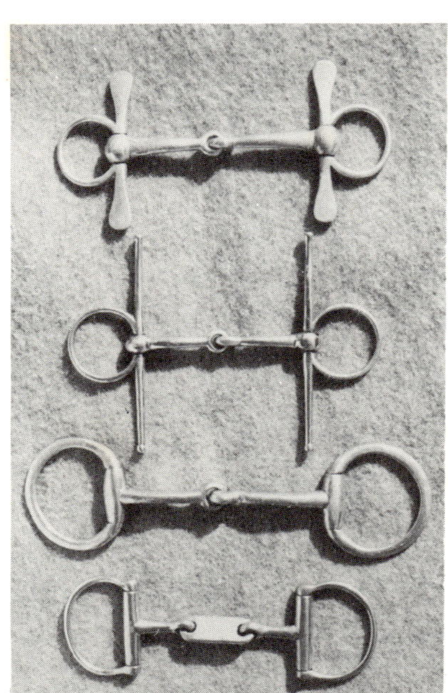

Fig 190. Jointed snaffles with and without cheek pieces

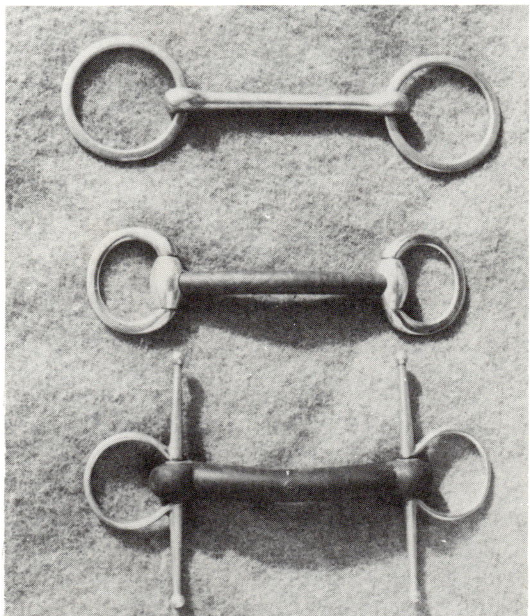

Fig 191. A–Straight bar snaffle
B–Half-moon egg butt snaffle
C–Half-moon cheek snaffle

above the mouthpiece determined the poll pressure (some horses flex better with more poll pressure), and the length below the bit gave pressure on the jaw in connection with the curb chain, known as curb pressure. Here I think we should consider curb chains because they can be very mild or severe. I do not feel enough attention is given to them—I cannot remember having been told about them in detail but I just found out about them for myself.

The mildest one is made of broad elastic with two or three rings on each end to adjust in size and one ring in the middle for the lip strap. The next mildest is made of leather in exactly the same way. There are rubber covers which can be slipped over a curb chain so that it lies more comfortably in the chin groove. There are many kinds of curb chains: the least severe are the large smooth links and the smaller the link the more severe the pressure—if you should have trouble in stopping, the old idea was to twist the curb chain several times. There are several very decorative double-linked chains used for showing. All curb chains should be worn with a lip strap, which stops the horse being able to take the end of the curb cheekpiece into his mouth. The port of the curb or Pelham bit can press into the roof of the mouth and as a result one can usually expect more flexion if it is unusually high. The less the angle in the port, i.e. the wider the port, the more room there is for the tongue. Many years ago Mrs de Salles le Terraire told me that horses with Arab blood in their veins usually have large tongues because coming originally from the desert they carried more saliva than a horse bred in colder climates. Therefore the larger tongue needs special consideration, usually a half-moon curb with a mild port gives the most room. Should a horse loll his tongue out to one side it will be found that given more tongue room he will hold his tongue correctly.

Having considered the general idea of the severity of the bits on the market, it might be as well to go into the action of the bit or the different actions that the ordinary snaffle, Pelham and double bridle have on the horse's mouth.

The snaffle bit tends to raise the horse's head. The majority of horses are inclined to lower their heads when asked to carry the weight of the rider as well as their own weight, hence the snaffle is the accepted bit for young horses in initial training. This tendency of the low head carriage is most uncomfortable for a rider and also puts a horse on his forehand, so that by raising his head the horse learns to balance his own weight and that of the rider in a better position. The snaffle is, I think, the most common bit in all the horse world, in fact no other bit is really used in racing. In hunting, show jumping, Pony Club and Riding Club work I feel the majority of horses

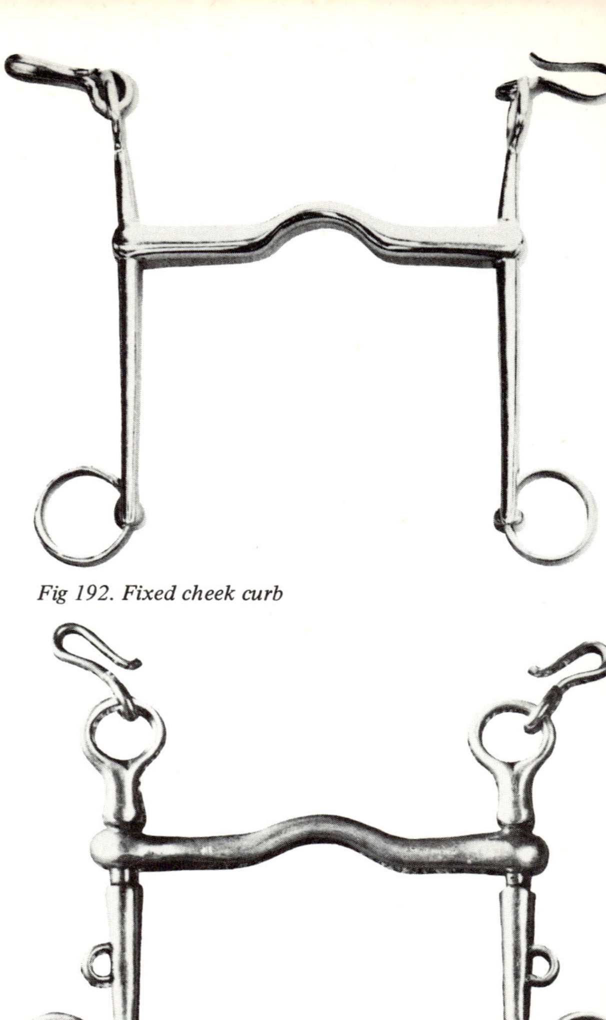

Fig 192. Fixed cheek curb

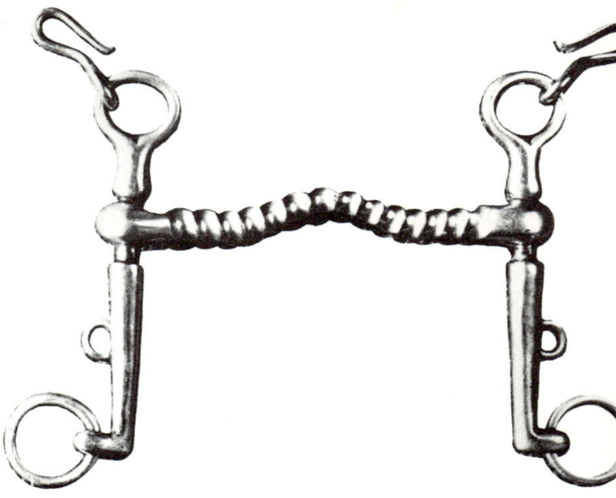

Fig 193. Ward Union curb showing smooth and serrated sides

Double Bridles

Fig 194. Fixed curb

Fig 195. Ward Union curb

Fig 196. Half-moon curb

will be found wearing snaffle bridles.

The double bridle has two bits in a horse's mouth–the snaffle or bridoon raises the horse's head and the curb lowers the horse's head (see page 21). By judicial use of both reins a good rider can position his horse's head exactly where he wants it, the horse becomes lighter in hand and the rider can obtain better control so that the horse can answer his wishes more easily. This can be seen in the show ring and in dressage tests. The two reins have different functions to perform and should seldom, if ever, be used with the same tension. In between the wars the curb with the sliding mouth-piece was generally used and was known then as the Ward Union. The horse can move this bit in his mouth and obtain relief from a constant pressure on one part of his tongue. In my mind I have always thought of it as a bit connected with hunting and the famous Irish Pack, the Ward Union. The other type of curb, the fixed cheekpiece or Weymouth bit, was difficult to obtain for some years as there was not much demand for it when only people of the 'old school' liked to use it for showing hacks. We considered it gave a horse a steadier head carriage

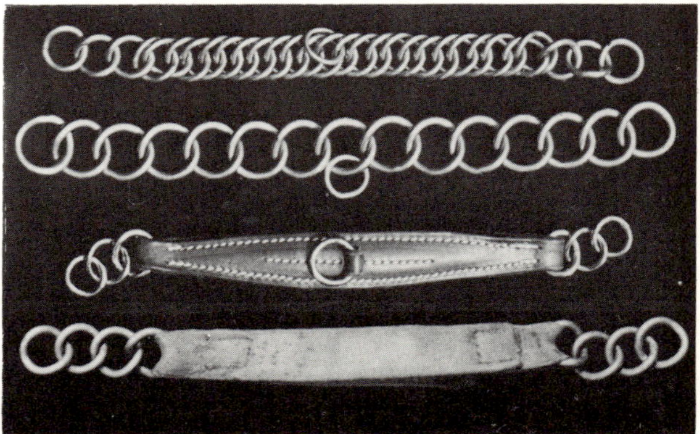

Fig 197. Different types of curb chains

and one did not lose him as he changed the position of his bit just at the wrong moment. Now this type of curb bit is much favoured for dressage. It seems that bits have fashions rather like ladies' clothes, they come in and go out.

The main function of the much despised Pelham is to lower the horse's head. One often sees a horse with his head too low, over bent, and behind the perpendicular in this bit. One can argue that it is one bit trying to do the work of two but it can be most useful when a horse is fussy with two bits in his mouth and is too strong in a snaffle. Again it has its uses should you wish to show or jump a horse that carries its head too high and star gazes. There are many variations of Pelham, and two kinds that I personally have found most useful are the Kimblewick and the Scamperdale. The former can be used with one rein and seems to touch the bars of a horse's mouth in a different place from a snaffle. Also, having only one rein it is most useful for children and gives the little extra stopping power that they cannot get with a snaffle. It was named by Phil Oliver after his farm, but it was originally brought to this country by the Spanish jumping team and is also known as the Spanish Jumping bit.

Another Pelham carrying the name of a farm is the Scamperdale, which has bent back sides and gives more pressure on the tongue: horses either like it or dislike it. We had three horses that went particularly well in a vulcanite version. In the case of *Tara*, flexion in other bits seemed almost impossible. *Tommy Tittlemouse* and *Alexander V* fought and were very difficult to control in other bits but dropped their heads and became much more manageable for jumping and cross-country work in this bit, invented by that great horseman, Sam Marsh. The official attitude is that the Pelham is not a good bit, I think because it is not accepted by the F.E.I. (Federation Equestre Internationale) for dressage but it has many opponents and one of the greatest was Robert Orssich. One day at a show when my daughter was very small and riding *Tara*, Robert said to me 'Why do you use that awful bit–Sam is not judging' and my reply was 'The mare likes it'. Robert said he was sure that no horse would ever go in such a bit but as the mare trotted past us in the most elegant fashion, well flexed and using her toe, he admitted 'So it works then'. And I would say when it works, use it, it can be an answer to a difficult mouth. But like all bits, do not expect that every horse will go in the one you yourself like: remember, you are not wearing it.

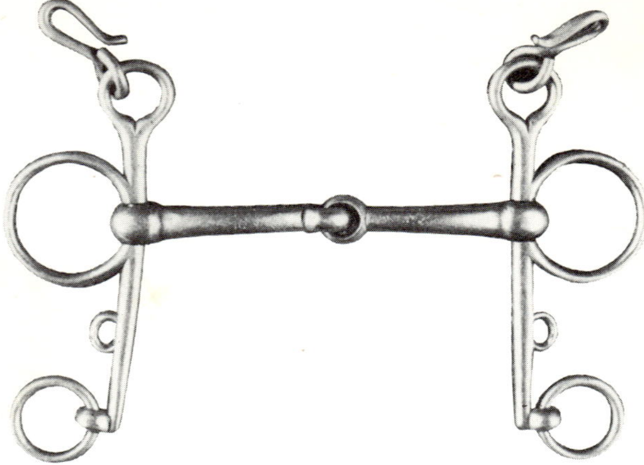

Fig 198. A jointed Pelham

Fig 199. A Universal reversible bit

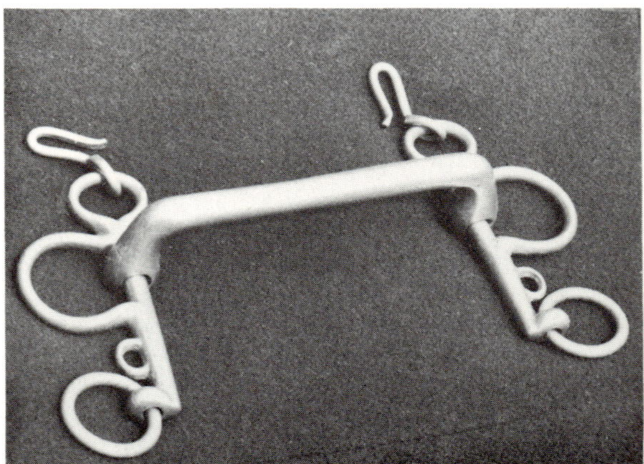

Fig 200. Scamperdale bit

Pelham Bridles

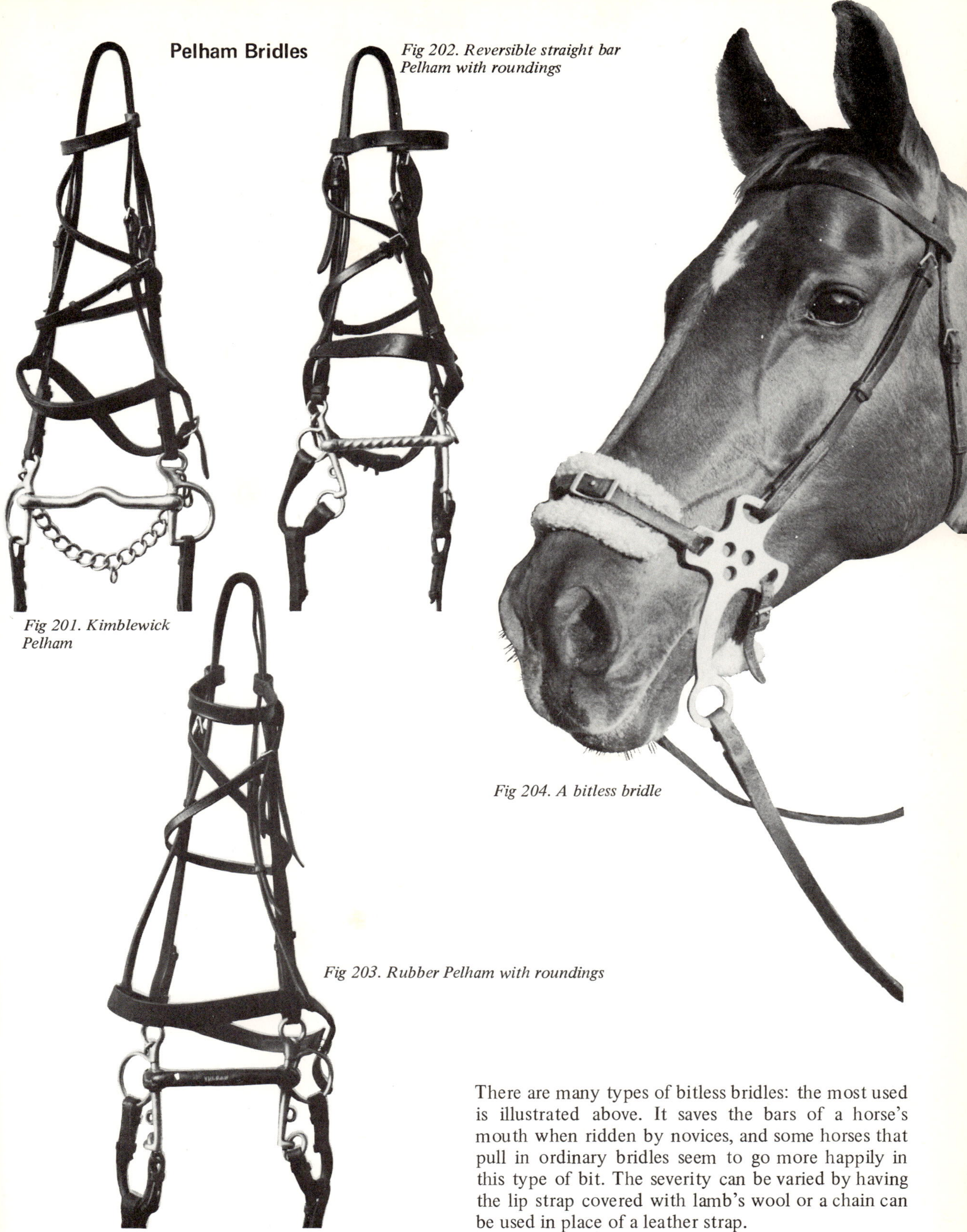

Fig 201. *Kimblewick Pelham*

Fig 202. Reversible straight bar Pelham with roundings

Fig 203. Rubber Pelham with roundings

Fig 204. A bitless bridle

There are many types of bitless bridles: the most used is illustrated above. It saves the bars of a horse's mouth when ridden by novices, and some horses that pull in ordinary bridles seem to go more happily in this type of bit. The severity can be varied by having the lip strap covered with lamb's wool or a chain can be used in place of a leather strap.

SADDLERY

All riders should be taught something about the tack they use when riding a horse; how the bridle fits and what it is made of, the necessary care of the leather, how to keep it soft for the comfort of the horse and to keep it from getting hard and brittle and unsafe to take the weight of the rider in the stirrup leather or the pull of the horse on the reins. If either of these vital pieces of saddlery breaks, the rider can be in danger of hurting himself.

The Saddle is the most expensive item of all the necessary things needed for the riding horse, but it can make or mar your comfort when riding and, for that matter, your position on the horse. First of all the saddle must fit the horse, as a badly fitting saddle can not only give a horse a sore back but can also roll in a most uncomfortable, or even dangerous, way on a round-backed animal.

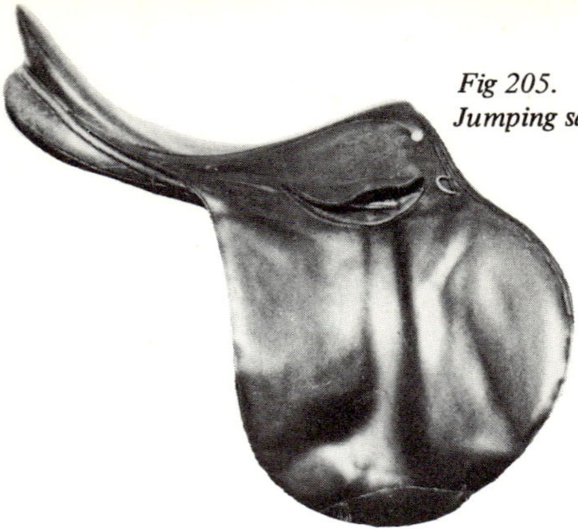

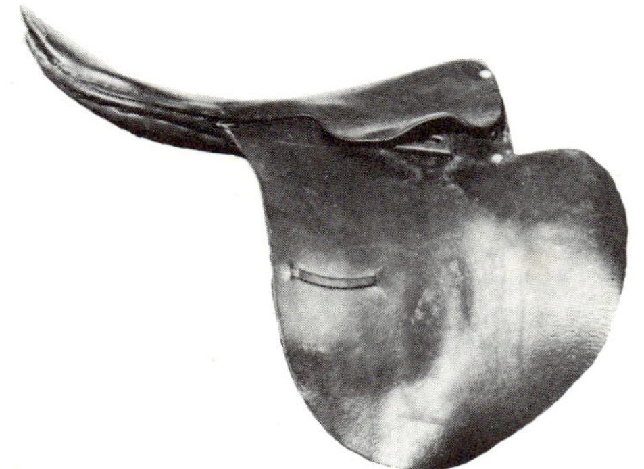

Fig 207. Racing saddle

Nobody expects the casual rider to know the correct names of each part of the saddle but at least he can have a rough idea of how it is made and, for his own sake, where to sit in it for his own comfort, i.e. in the lowest part of the saddle, see Fig 2. For the comfort of the horse he should ensure that the padding on the under surface is sufficient to keep the weight of the rider from pressing the saddle down on to the spine of the horse. The rider should be able to put two fingers between its front arch and the horse's backbone, and if he looks down he should see daylight right through the channel of the saddle–this works in two ways, first as ventilation and second to keep the pressure off the spinal column.

There are many types of saddles made; each has its own function, to make the rider comfortable and to make the job the horse is asked to do easier for him to perform, so it is up to the rider to buy the kind of saddle he or she needs.

When doing specialised work we need a specific type of saddle, i.e. a *Racing Saddle* for point-to-

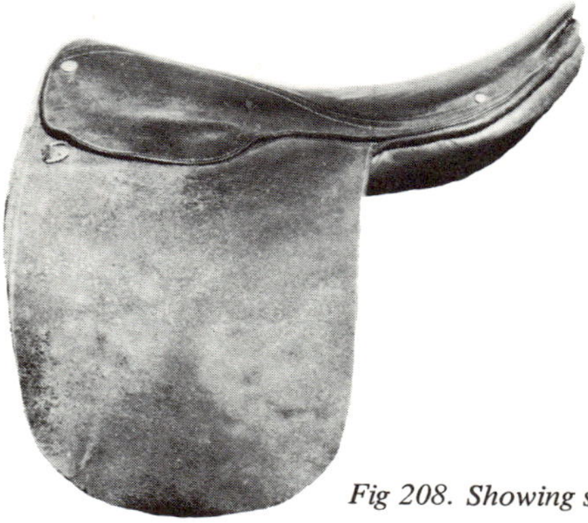

Fig 205. Jumping saddle

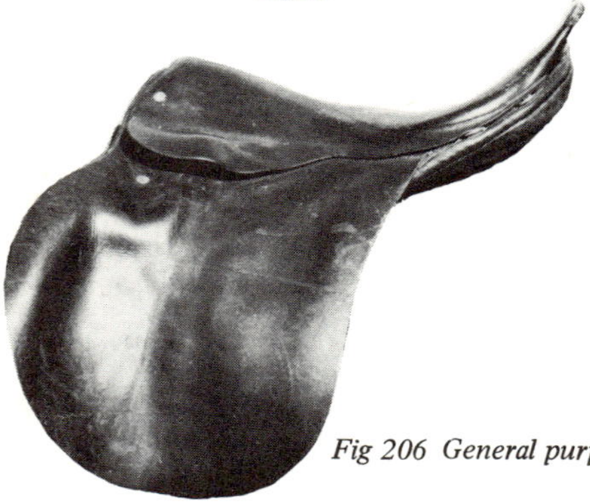

Fig 206 General purpose sad

Fig 208. Showing saddle

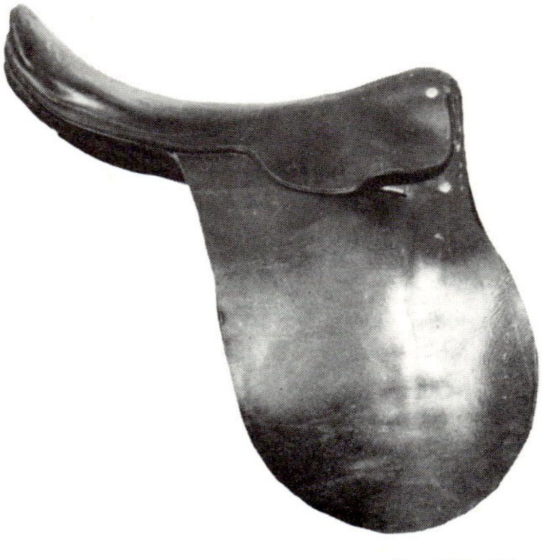

Fig 209. Hunting saddle

Fig 210. A well organised tack room!

pointing, a *Straight Cut Saddle* for showing. A *Jumping Saddle* can be any that is comfortable and centrally seated. It is nice, of course, to have a special *Show Jumping Saddle,* but on the whole not really necessary as long as it is not a saddle which tends to throw the rider's weight backwards.

There is a saddle which is now produced called the *General Purpose Saddle.* This, as its name implies, can be used for most things. It has a deep seat and is forward cut, so it is comfortable for the rider and enables him to get forward into the knee rolls for jumping and cross country work. Most of the modern *General Purpose Saddles* have spring trees and are very soft, being padded with foam rubber. It can be used in working hunter classes where performance and a good ride are more important than conformation. There are, of course, corresponding saddles made for children.

A saddle is like a good shoe–it needs fitting and requires care and cleaning to maintain the leather in pristine condition. Leather has *two* sides–a grain side which is the outside of the skin from which the hair has been removed and a flesh side. Each requires a different treatment. All saddlery and harness leather has a fat content: its loss, during wear, must be replaced if the leather is to remain supple and long lasting.

There are a number of excellent preparations on the market for feeding the leather.

1 The underneath (flesh side) of the saddle flaps, girth straps and panels should be cleaned of sweat and salt after use with luke warm, NOT HOT, water finishing with a piece of oily (neatsfoot oil is especially good for this purpose) hessian. **Never** use vegetable or mineral oils. Panels should be well brushed to remove loose hair and dirt. Be careful to dry all metal parts to avoid rust.

2 A leather feeding preparation should be well rubbed in to the flesh side of the leather at least once a week.

3 The grain side should be well massaged with a sponge or rag using saddle soap without water about once a month.

4 The saddle, when not in use, should be placed on a rack to allow air to penetrate under the panels.

5 Always dry a wet or damp saddle in natural air-temperatures. NEVER hang a wet or a damp saddle over radiators etc.

6 Always store saddlery and harness in a dry room.

Stirrup Leathers

The choice of stirrup leathers is optional; on the whole I feel it is better to buy the most expensive raw hide which are more or less unbreakable, so they are safe as well as being soft on the shin of the rider's leg. The stirrup irons are important in size and material. Again the most expensive stainless steel are best—they are safe and do not bend or break, as nickel does in time. They must fit the rider's foot so that, should he fall, his foot is not caught in too small an iron so that he gets dragged, or slip as a result of having too big an iron. I would advise the usual bent iron, since it neither presses against the instep if the rider slips his foot home in the stirrup, nor rubs his hunting boot. *The Kournakoff iron* is a slanted iron, both from the side and on the tread of the iron itself, so that the iron hangs to one side easing the rider's foot inwards to take the weight correctly on the big toe joint and, at the same time, slanting the foot with the weight to the rider's heel. In other words, it is wonderful for people who have difficulty in keeping a good foot position, but put on the wrong way they are most uncomfortable and do the complete opposite from being helpful. Again being the most expensive, they are probably worth it in the end. Most people now like rubber treads put into their irons: these are non-slip and help to keep the foot in position.

There are *safety irons* for children on the market, but for the most part they are not very safe: I have seen a peculiar accident with a Peacock safety iron, in which the shoe-lace got caught up on the hook for the india rubber band and the child got dragged. The only really safe stirrup for children is the old fashioned clog. These are not popular with the young as it is felt they are not real stirrup irons, and children long to wear grown-up irons. My brother and I rode in them and my daughter used them for jumping, even at the Royal International when she was ten or eleven years old. I think they do two things for the young rider—the child can never get dragged and he has to keep the weight on the ball of the foot.

Girths

These are usually made of nylon, leather or lamp-wick. There are many kinds of girth on the market, and the general attitude has changed very greatly since I was young, when those we all used were very wide, as shown in the photographs of leather three-fold girths used for side-saddles. Now we use narrower girths, and have almost discarded the leather folded girth as being cumbersome and impractical since, unless kept really well oiled, it tends to rub and give girth galls. The lampwick girth has become popular of late. Nylon or string girths have a good reputation, but do not wear very well and if the animal is fat, unless the leg is pulled well forward, it is inclined to pinch or cut in behind the elbow. The leather Balding girth is very good as it is so shaped that it is narrow behind the elbow; it must, however, be kept well oiled.

Fig 215. Girths

Fig 211. Kournakoff

Fig 212. Bent iron

Fig 213. Peacock safety iron

Fig 214. A Clog Stirrup

Bridles

There are two main kinds of bridles: Single rein, which have come to mean snaffles but can be single-reined Pelhams, or double rein bridles. Whichever is used, the bridle consists of different pieces. Cheekpieces and reins can be sewn, studded or billetted on to the bit. The *headpiece* is made of leather and goes over the head, having straps which can be adjusted on to the *cheekpieces* i.e. one for each side to hold the bit in the horse's mouth. The *Browband* fits round the horse's brow and can be of plain leather or fancy or coloured. It should be long enough not to pull the headpiece on to the back of the horse's ears, which is uncomfortable and often makes a horse throw his head about. The *noseband* is fitted two fingers' length below the cheek-bone; it can be narrow, wide or stitched most ornamentally. It is useful to attach a standing martingale on to and, as can be seen from prints, was not worn very much last century. When a double bridle is used there has to be a bridoon bridle headpiece which slips between the headpiece and the noseband through the browband, and holds the bridoon bit in the horse's mouth.

Reins can be very varied in width, length and kind. The ordinary rein is usually $\frac{7}{8}$ inch wide to be both serviceable and tough, but most of we women like more delicate reins to handle. One I particularly favour is a narrow plaited rein specially made for me by Mr. Kimble, my saddler for 30 years. It measures ½in. in width and is made to whatever length we ask for. There are rubber-covered reins for racing or horses that pull; plaited or laced reins to help a rider whose horses take a firm hold; variations of nylon and string reins are used a great deal for jumping since they are less likely to slip in the rider's hands, even in wet weather. When using double reins, the rider should see that the curb rein is slightly narrower than the bridoon rein, or it can be a finely plaited rein.

In-hand showing-bridles have been very well developed to meet the growing demand, with so many classes for all types of ponies and horses in breeding classes. These bridles seem to be mostly variations of the old type stallion bridle, and I must say some neat little Welsh mare heads look most attractive with their narrow brass browbands and brass rosettes by their ears. Years ago this type of bridle would never have been used except on a stallion. There are also some delicate in-hand bridles for all sizes of ponies, and the coloured browband is much favoured.

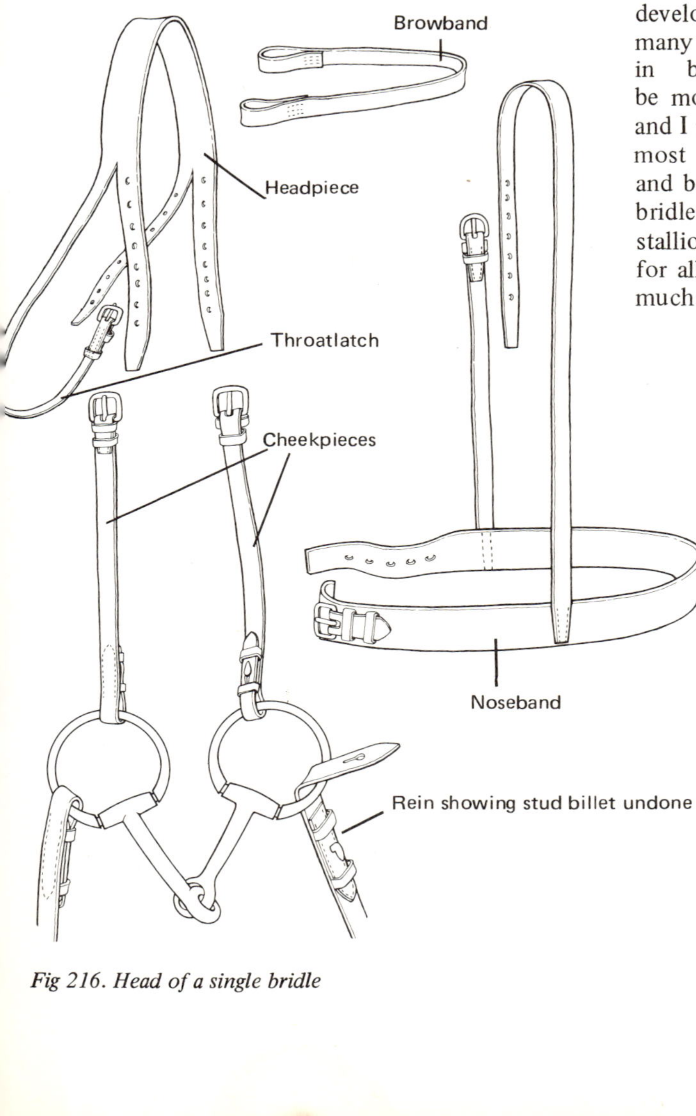

Fig 216. Head of a single bridle

Fig 217. Laced and plaited reins

Headcollars

Headcollars are made of leather, webbing or nylon. There are some very pretty brass-fitted ones, or cheaper string versions with tin furniture (as the buckles and dees are called). The best for rough everyday use are those made of rawhide which are soft and do not rub a horse's head. They do not go hard if lost in the field for several weeks, they need no cleaning and are, to my mind, the most practical on the market, although not attractive to look at as they are rivetted and made with tin furniture.

Clothing

To keep a horse warm one must have rugs—there are many types and all have different purposes. The most common in any stable is the *jute night rug* which can be half lined or fully lined with blanketing. When half lined with its own surcingle, this is known as a *dealer's rug,* since it needs no roller to keep it in place, is less expensive than a fully-lined night rug but is sufficiently warm to keep a horse from getting cold. It is often bought by dealers at a sale, hence its name. A fully-lined night rug is the type found in general use in most stables: it can have one or two surcingles on it, or none, in which case a roller has to be used. The closer the weave of the jute the more hard-wearing the rug, but the best of all (and, of course, the most expensive) are the flax rugs.

All rugs are made in different sizes and can vary by 3 inches from 4ft 6ins for a small pony to 6ft 9ins for a larger horse.

Again for warmth there are *blankets,* usually beige with red and black stripes running each side of the rug. I think this pattern has been used through the ages to help grooms to put the rug on evenly. Blankets vary in size and in weight, the heavier ones weighing 8lbs down to the lightest which weigh about 3lbs. Here, I think one should mention that, with present-day prices, one can substitute an Army blanket, which is pure wool and much lighter, but costs £1.50 as opposed to £7.00. They tend to be very long but fold back well and can be tucked under the roller to secure them.

When a horse is brought in hot, an *anti-sweat sheet*—i.e. one made of material on the lines of a string vest—is very useful to put under a light rug to dry the horse off. Or the old-fashioned cooler, made with a browband and a fitted neck piece, and hanging loosely over the horse, allows air to circulate. It should be of fine wool, or one can make one's own out of an Army blanket, since the length is there and it is easy to shape the neck.

The nicest rugs of all are *day rugs*, which can be obtained in very attractive colours with contrasting bindings. The most expensive are livery cloth, through which any wind would find it difficult to penetrate. Recently it has become the accepted practice to use a lighter and more porous cloth for day rugs; even so, they can vary a great deal and some summer showsheets are very light in weight and quite unsuitable for keeping a horse warm in winter. Next

Fig 218. A day rug

Fig 219. A blanket

there is the *paddock sheet*, a small rug used for horses parading before a race. As they are worn over the saddle, these sheets are very useful for exercising in the winter, just to keep the horse warm over his quarters, not too cumbersome or long enough to get splashed and dirty when trotting on wet roads.

Most stables choose a colour and have an initial put in the corner. Nothing looks smarter than a horse, rugged up with matching knee pads, tail guard and bandages, being unboxed at a show.

Lastly there are *summer sheets,* which can be made of cotton or linen, to be used in the summer months on hot days. They used to be known as *fly sheets.* I think aptly named as they keep the flies from settling on stabled horses and annoying them.

Rollers are used to keep a horse's rugs in place. They can be made of leather or webbing and there are many designs. One of the most useful is a small anti-cast roller made with straps either side, so that any length of girth can be used, thus enabling the roller to be used for either a pony or a large hunter. Breast girths can be attached to any roller to help hold it in place on a horse who is inclined to have no spring of ribs.

Fig 220. Different types of rollers

Bandages are a necessity in any stable, however small. In case of illness woollen ones will keep the extremities warm; they are used for preventing such things as cracked heels in winter, besides being used over Gamgee for travelling. Stable bandages are usually grey and made of light material–the coloured ones are usually thicker. There are stockinette bandages, linen bandages and elasticated and tail bandages. Again these can be obtained in many colours. All leg bandages, in my opinion, should be used over Gamgee or cotton wool to avoid bandaging too tightly and interfering with a horse's circulation.

Tail-guards can be made of leather and be quite elaborate in design, or they can be made of rugging or jute and are put on over a tail bandage to 'guard' the tail when travelling in case a horse sits back on his tail to balance himself. They can also be used in the stable should a horse have the annoying trick of rubbing his tail (of course this could indicate worms, but some just do it, even when there are no worms present to irritate them).

Knee-caps are to protect the knees of a valuable show horse when travelling or exercising on the road. There are many kinds and they can be made of leather, or rugging or jute. Leather last longest and are best in the end, although expensive. I prefer the very light kind, known as skeleton knee pads since they can be used for both purposes.

Fig 221.Skeleton and ordinary knee pads

STABLE MANAGEMENT

Stable management, or the care of a horse, be it at grass or kept in a stable, is most important, because if the horse is not well in himself he cannot work well nor give his rider the pleasure he expects.

First we will take the horse or pony kept at grass and we will make one maxim: 'Man cannot live on bread alone—nor can horses live on grass alone if we expect anything more from them than about 1 hour's work of quiet hacking'—and even then, except in summer, he must have supplementary feeding to make up for the loss of nutrition in the grass after July. The horse kept out should have ideally at least one feed per day with as much hay as he needs. In the winter, a half-bred horse can live very well in a New Zealand rug, with adequate supplementary feeding, and can do his weekend rider credit at local shows or give him a day's hunting on Saturdays. But we do not expect these horses to hunt hard nor event nor race.

When a horse is kept in a stable, he is kept under unnatural conditions to satisfy the ego of his owner, who wishes to hunt, race, show jump or show. His care is somewhat time-consuming because he needs regular feeding and regular exercise. This is almost impossible for someone who is working all day and has nobody who can give the horse a feed in the middle of the day. When a horse is confined in four walls, the main object must be to keep him happy and amused: by daily exercise and training, the owner helps to break up the long hours of enforced boredom. Yard diversions such as pigeons, bantams, cats, dogs or even goats give horses something to occupy their minds, although these pets do not help the tidiness of any yard!

In a large yard there should be lists of feeds written up to help anyone who is not the usual feeder. All feeds should vary a little each day according to the horse's needs. The degree of dampness will vary according to the condition of his stomach. The amount of oats or energising feed should be increased or decreased according to his work. Over-feeding can be far more dangerous than under-feeding, although continued under-feeding can lead to many troubles. By over-feeding, I do not mean the giving of too many oats, which makes a horse play up and be above himself; I mean the continual over-loading of his system, which can cause swollen legs or the more serious Azoturia which can cause death if not dealt with in time. This complaint seems to be more prevalent in recent years. I feel it could be caused by the excessive use of artificial manures on farms to produce heavier crops, thus causing the imbalance of minerals in hay and corn, which could act on the horses internally, producing deficiencies in their feed. Feeding is therefore not quite as straightforward as it used to be when good hay, oats and bran seemed to produce all that one wanted for horses in work or breeding stock. Now we have to

Fig 222. The stable yard

use additives in feeds to make up for the deficiencies in minerals such as magnesium, caused by using artificially produced food. The bulk content of a horse's feed is made up of hay and I have often wondered how best to explain this. I quote now from a list which hangs in my hay shed and which reads thus: 'A horse has hay plus four feeds of corn per day, hay nets contain roughly:-

horses between 18 and 22 lbs a day to get fat,
horses already fat 14 to 18 lbs per day,
horses that are too fat or have Laminitis 8 to 14 lbs per day according to size'.

Then there is a list of horses and beside each name appears the weight of hay that should be given. There is a spring-balance hanging in the hay shed so that each net can be weighed. When two weights are given on the chart, it refers to nets given in the a.m. and the p.m. for those horses who are greedy and eat all their hay before evening stables. When one weight appears, it means that a horse gets one haynet for the 24 hours. All nets are put up before breakfast, and any short feeds which have been left are reported, i.e. the person in charge of the yard and feeding has some idea if the horse is off his feed and should be looked at. As for short feeding, again there is the list in the corn shed and the feeds vary tremendously. Point-to-point horses get between 16 and 20 lbs of concentrate per day. Hunters between 10 and 14, riding school horses according to their work, children's ponies may have as little as 2 lbs of bran divided into four feeds with hay chaff added, so that each horse is fed according to its work. As a yard, we are great believers in the use of good hay chaff to stop a horse bolting his food and not getting the full value from it.

Daily Routine

7 a.m. short feed, muck out and water.
8 a.m. hay up and water.
8.30 a.m. breakfast.
9.15 a.m. tack-up for exercise and schooling.
10.30 a.m. tea break.
strap
12.00 feed, clean tack.
1.00 p.m. lunch.
2.00 p.m. skip round.
2.15 p.m. schooling (2nd period).
3.30 p.m. do up stables.
4.00 p.m. feed.
4.30 p.m. tea.
7.00 p.m. late feed, skip round, check water and hay.

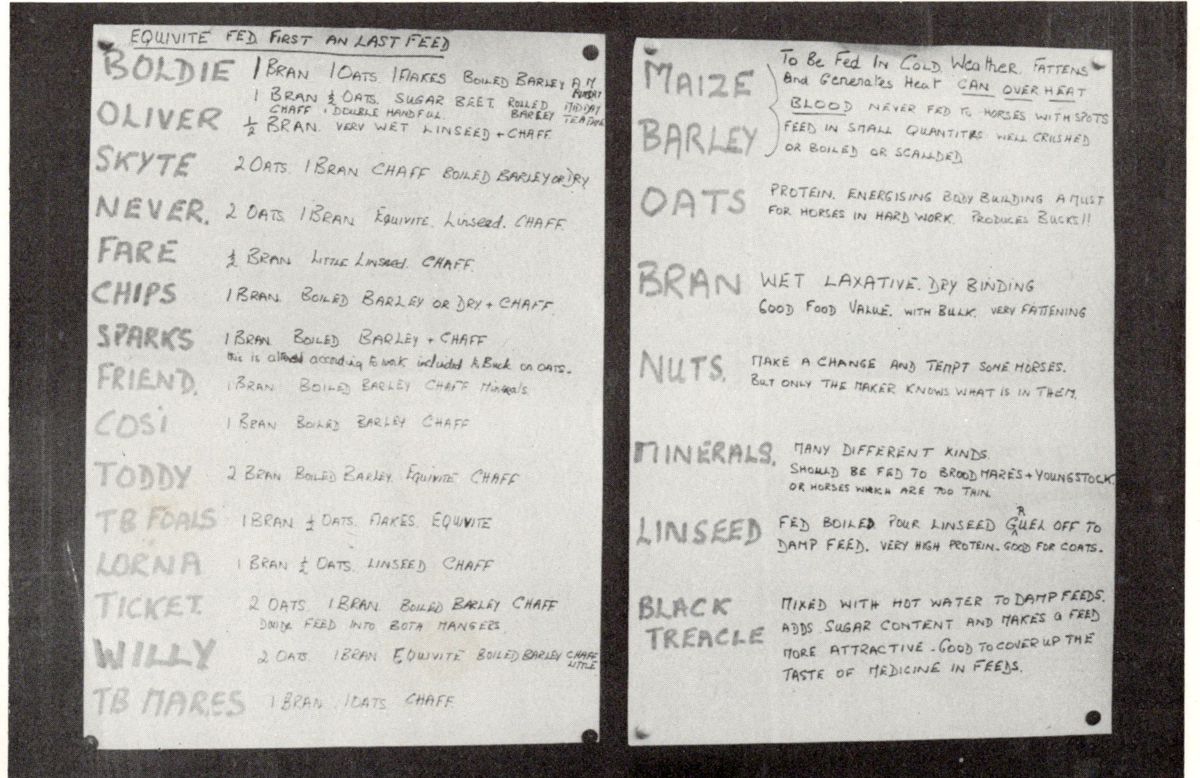

Fig 223. Feeding instructions pinned up in the corn shed

FOOD

Oats—have been the accepted horse feed over the ages. Horses like it and it gives them energy to cope with the hard work demanded of many of them—it is not so suitable to feed in any quantity to children's ponies. Oats can be crushed, and they are then suitable for young stock and old animals whose teeth are not so good. Rolled oats are more often used for cows but can be fed to horses. Cracked oats are used for horses with good teeth. 'Cracked' means that the husks are just split and the oat is mostly left inside its husk.

Bran—is the next most important feed: horses like it and it is the one commodity really used in sickness and in health. Fed wet it is laxative, fed dry it is binding, mixed with other short feed materials it makes good bulk, is palatable, is easily digested, and does not make horses silly in their behaviour. In fact, if I were given the choice of only one feeding material, it would be bran.

Flaked Maize—is imported from abroad. It is very good for fattening. It is a warming feed for the winter months, and should be fed in small quantities dry, or scalded with boiling water, and can be mixed with other feeds. It is not so good in summer, when it can cause over-heated blood.

Barley—may be fed boiled or well rolled, as it is inclined to swell if eaten whole and could give colic. It makes an excellent hot feed when boiled and added to bran and oats.

Linseed—is very high in protein and when boiled is used to damp feeds. Horses love it and it has great medicinal properties. To make linseed tea, put one teacup full in 2 pints of water, bring to the boil and allow to simmer (overnight, if possible).

Sugar beet pulp—has a high sugar content, adds bulk to short feeds and helps to fatten horses. It must be soaked for at least 8 hours before use and should be used during the next 24 hours. It is dangerous to use undersoaked or stale as it goes sour very easily and can cause severe colic.

Nuts—can be made from any grain or mixtures of grain, ground and mixed with molasses to bind them. They can be made also from grass. When made by a reputable firm they are safe to feed, and nothing need be added for ponies since they are a balanced ration in themselves. They are good to add to feeds to tempt shy feeders, as most horses like them.

Milk equivalent or calf powdered milk—this is excellent for putting on condition: it has high protein value and is very useful for stock which is not doing well. Excellent for newly foaled brood mares who are lacking in milk, and for foals when weaned. Most horses will eat it if introduced in small quantities into their feed, working up to one handful per feed, i.e. about the equivalent of 1 to 2 gallons of milk per day. Some horses will drink this from a bucket on returning from trekking or hunting, which is excellent.

Carrots, turnips, apples, etc.—when cut up and added to feeds, horses always enjoy them as a change.

Potatoes—are not supposed to be good for horses but they are fed a lot in Ireland, either raw or boiled.

Hay—Since this is the bulk food of all horses which are stabled, it is of the utmost importance to see that it is of good quality. There are three main types of hay:

1 Meadow hay, which is soft and contains a variety of grasses, and is more palatable than most. It is very suitable for young stock and brood mares, since it contains a more varied assortment of minerals from each kind of grass. It is also good for old horses whose teeth are no good for hard hay.

2 Seed hay, which is hard hay, is made from a grass grown from "seeds" sown by the farmer. It is usually Rye grass and Clover—a one year's lay, very high in protein. It is a stalky hay and rather too hard for young stock, but excellent for horses in hard work such as hunting and racing.

3 Lucerne hay when made correctly, is the best of all for horses, but it is very difficult to make since the stalks contain a lot of sap and the leaf is very easily lost if too dry. If baled slightly wet, it very easily becomes musty.

All hay should be sweet smelling, a light brown or greenish in colour if meadow hay or seed hay, while Lucerne should remain its greenish colour. Any hay that becomes too dry with being stored should be well shaken out and damped by dipping the hay net into water and leaving it to drain.

Fig 224. A hay net too low is unsafe

STABLE VICES

Fig 225. Bars across the stable door to prevent weaving

Most vices or bad habits in the stables are caused by intense boredom. When a horse or pony is kept constantly in a confined space, either because he has hurt himself, or his owner has not enough time to ride him regularly or properly, he does not get enough exercise. It is not surprising that he thinks up some game to relieve his boredom, rather as a child, if left in bed for hours on end, will pick the wallpaper off the wall or play Os and Xs on it, or tear his sheets, just to have something to do. Some horses have pretty harmless but annoying tricks like upsetting their water buckets, scraping their beds about or merely standing scraping the floor of their box. These are annoying to the groom but not of great consequence, except if the horse scrapes with one foot he wears out one shoe faster than the other three. *Kicking the boards or the water bucket* is an annoying habit which the other horses copy. The cure for both of these is to deaden the sound. Pad the boards with an old mattress or cover the walls with sacks stuffed with straw. But creosote them first in case they are torn for amusement. Put in a plastic water container that does not rattle when kicked. To relieve boredom and give the horse a change, try to turn him out in a New Zealand rug daily.

Although the above helps with the next two vices, it does not always eradicate them. A horse will weave over a gate in the field or crib bite on a fence or wind suck. *Weaving* means a horse swings from side to side, putting his weight first on one foot then on the other. Weaving is a habit which lots of Thoroughbreds develop. Some horses do not lose condition nor do they seem to put a strain on their front legs, but a weaver is something nobody likes to admit having in the yard. It is a habit that other horses follow and it constitutes an unsoundness if the horse is to be sold. The old-fashioned inside stables are good in such cases as there are no doors over which to lean. This weaving can be stopped to a great extent by dissuading a horse from rocking over his door. If you have not got inside stables you can have bars put on the door, wide enough apart so that the horse can look out but not wide enough for him to swing without bumping himself. The bars in the photograph are 12ins. apart. Care must be taken not to open the door until the horse's head has been pulled back.

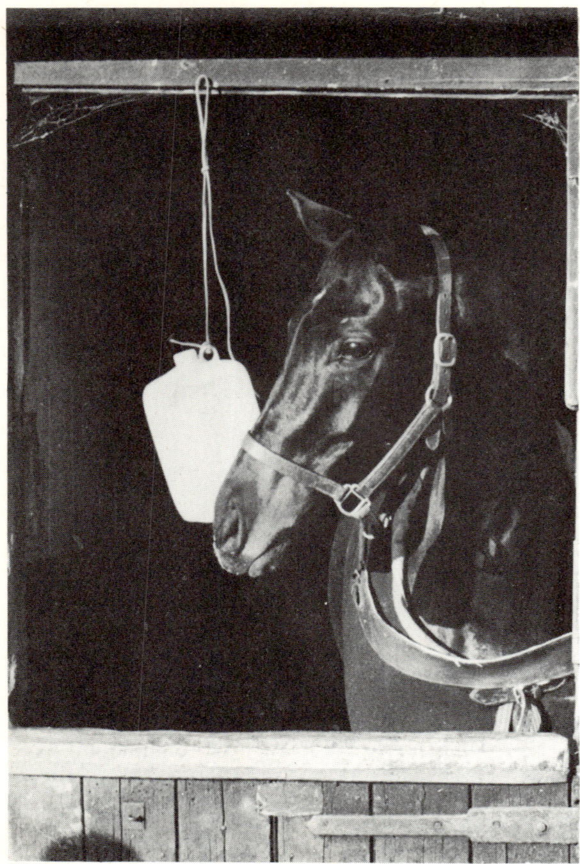

Fig 226. A plastic orangeade container

Fig 227. Crib biting strap

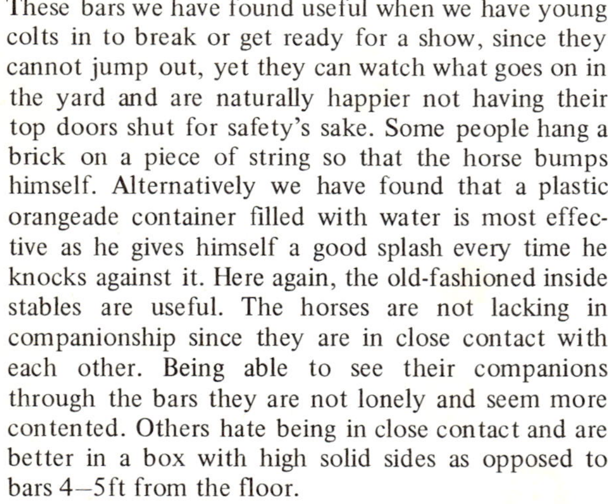

These bars we have found useful when we have young colts in to break or get ready for a show, since they cannot jump out, yet they can watch what goes on in the yard and are naturally happier not having their top doors shut for safety's sake. Some people hang a brick on a piece of string so that the horse bumps himself. Alternatively we have found that a plastic orangeade container filled with water is most effective as he gives himself a good splash every time he knocks against it. Here again, the old-fashioned inside stables are useful. The horses are not lacking in companionship since they are in close contact with each other. Being able to see their companions through the bars they are not lonely and seem more contented. Others hate being in close contact and are better in a box with high solid sides as opposed to bars 4–5ft from the floor.

Crib biting is when a horse opens his mouth and takes hold of the door or any projecting piece of wood or a manger with his teeth and usually they *wind suck* at the same time. Really bad wind suckers also do it when standing in their stables and not holding on to anything with their teeth, merely bending their necks and gulping down air. These horses usually do not carry any condition, and the wind gives them pot bellies and upsets their digestion. In some cases horses do not do this when turned out, though confirmed wind suckers will hang on to any post and indulge in this vice even worse in a field.

For those horses that will chew wood or crib bite, there are many different patent preparations on the market to stop them, Nonaro, Cribox, etc., but plain creosote is as good as anything and more wholesome. *Rug tearers* are also annoying but sometimes this is caused by a horse having an extra sensitive skin. The woollen rug tickles him so much that he goes mad. I can sympathise with this as woollen vests used to do the same to me at school. As with children, if a smooth material is put next to the skin the horse will find relief and stop tearing his clothing. A cotton bed sheet or a piece of linen material will suffice. But if this is merely a mischievous trick, a bib put on the head collar side dees and hanging below the lips stops this most destructive and expensive habit.

GROOMING

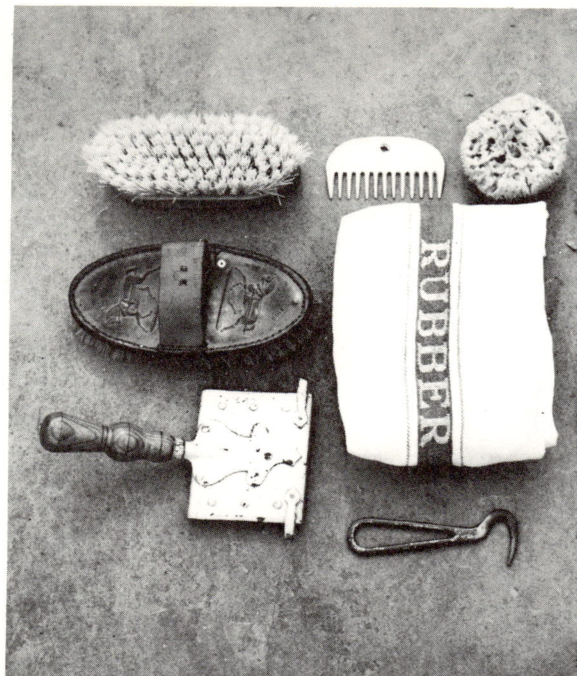

Fig 228. Grooming kit

Dandy Brush

This brush is the most widely used, both for stabled horses and for those kept rough in the field. It removes the dirt from the coat but should not be used hard enough to remove the natural grease, which is so essential to any horse which lives out without a rug and has to stand up to the wet and cold of the English climate. There are several kinds of bristles: the best is the light colour wire-drawn bristle, which has a nice give in it. The dark colour bass bristle used in the indestructible brushes is a harder type, never appreciated by the thinner-coated animals.

There are also nylon bristles but they do not seem to stand up to really tough wear. Recently there has been a very nice soft-bristled dandy brush produced which I have found excellent to use on thin-skinned horses who are either clipped out or kept in the stable.

Body Brush

There are many different kinds which are used really to clean and remove all dust and dirt from a stabled horse's coat. Again there are several types of bristles used: a fibre bristle can be of two different qualities, but the softest of all is the real hair brush, suitable for really fine coats. The prices of these brushes vary tremendously—the nylon brush is the cheapest and the leather-backed hair brush is the most expensive.

Curry Comb

The curry comb is used to clean the body brush. It can be of two main types, with a handle or with a piece of webbing across the back through which the hand is slipped. It should never be used to remove the mud from a pony's legs as it is too hard to go over the bones of the hocks or the knees etc. Recently there has been produced an excellent tough plastic appliance called Curry Flex. This is excellent for getting the dirt off muddy field ponies but the teeth are inclined to wear and they tend to tear the mane and tail unless used with the greatest of care.

Stable Rubber

This is used to polish the horse off and to wipe his head and ears when he has been groomed. It is usually made of linen, although some people prefer silk or chamois leather.

A wisp is made of hay and used to tone a horse's muscles up by rhythmically banging. This makes the muscle contract and relax, and is excellent for getting horses fit for racing and improving the neck and quarters of a show horse.

Sweat Scraper

Used when washing your horse down after strenuous work to remove superfluous sweat and water when he has been washed down, or when a grey or Palomino has been washed and rinsed, or when quartering in the morning. These can be made of pliable brass or they can have a handle, giving one side brass and the other side rubber. Should you not have the rubber one, then the groom's hands should be used to remove the water from the horse's legs or his head.

Usually a sponge is used to wipe out the eyes and dock of a horse, and here we come to a very useful piece of most inexpensive equipment, the *Irishman's Sponge*—cut a piece of sacking of convenient size for the hand, wring it out in hot water then it can be used for the eyes, nose and dock as any ordinary sponge. It is also used, in a circular movement, instead of a water brush to clean the horse's coat. It will be found most useful and will remove a lot of dirt if it is rinsed and wrung out each time a portion of the horse is cleaned. He should then be strapped dry with the body brush. Recently the introduction of Cactus Cloth for strapping has been much advertised. It certainly seems useful as dry it removes mud from a horse's coat and wet is used much as an Irishman's Sponge. It is, of course, much rougher in texture than sacking.

Water Brush

Used on stabled horses to help remove dirt and grease from the coat, should be used only just damp in a circular movement to get underneath the coat, and also on the mane and tail and to get straw marks off a stabled horse when quartering. Very useful when cleaning the stable stains off a grey horse in the morning.

Practical Hints on Grooming

Grooming for the horse kept at grass should be merely carried out with a dandy brush to remove the dirt which has accumulated in his coat from rolling and lying down in a field. It is essential that the natural oils remain in his coat to keep any rain from penetrating it and chilling him. Should you by any chance wish to show a horse then bring him in from the field and shampoo him. After this it is advisable to keep him in for at least two days to give time for the natural grease to return to his coat.

For the stabled horse, grooming is a kind of massage and the idea is to remove the natural grease from his coat and to give him a super, well-kept effect. Again a dandy brush is used to remove any superfluous dirt, followed by the use of a water brush, slightly damp, using a circular movement to disturb the dirt and finished off by going the same way as the coat. Next the body brush is used again in a circular movement to remove all the dust from the coat. To finish off the work of the body brush, one should lunge the weight of the person grooming on to the brush, working the way of the coat to produce a shine. The body brush must be cleaned at intervals by the curry comb and then the horse should be wiped over with a stable rubber. Care should be taken to pick out a stabled horse's feet twice a day to avoid thrush. The feet should also be oiled to maintain the natural growth and suppleness of the horn of the hoof, which is deprived of the dampness from the dew in the grass or on wet days. The mane and tail should be water brushed and body brushed daily.
When the horse has a short coat, either in the summer or when he has been recently clipped in the winter, the Irishman's sponge may well be used in place of the water brush.

The sponging out of the eyes, nose and dock should be done daily as well.

A Thoroughbred or show pony, with a very sensitive skin, may find a dandy brush too hard to accept gracefully. As stated, a very soft dandy brush has come on to the market and it is most useful for this type of animal. Before they were introduced, we always used a stiffer body brush than the ordinary soft hair one used on a good-coated animal.

Care of the Feet

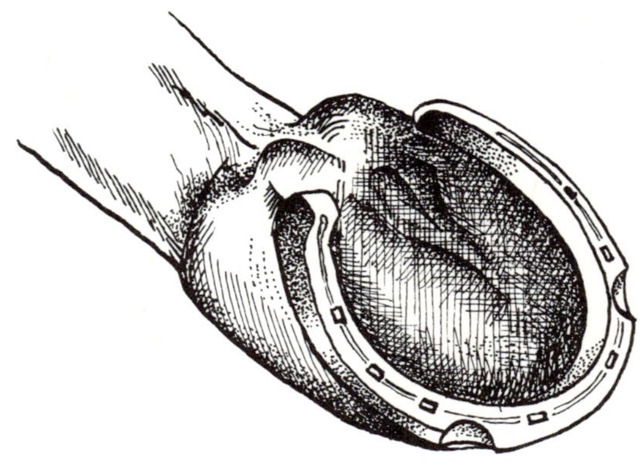

The old adage 'no foot, no horse' is as true today as it was 50 years ago. Another version is to look for the lameness of the shoulder in the foot. In other words the horse's mobility comes from the use of his feet and without sound feet he cannot be much use. The foot comes into contact with the ground, be it slippery clay, motor road or uneven rough and stony ground. Therefore it is the horse's foot which is most likely to get damaged or at least near the foot, the fetlock, will take the concussion of the foot. Stones can cut or bruise the sole or frog. The protection of the foot on hard going is therefore most important since the horse was not designed by nature to carry the rider's weight plus his own on hard surfaces such as the artificial tracks and roads of today.

This again brings to mind the rhyme, 'Tain't the 'unting on the 'ills that 'urts the 'orses' 'ooves, but the 'ammer, 'ammer, 'ammer on the 'ard, 'igh road what does the 'arm!' The banging of a spanking trot on a metalled road causes jarring and splints, ring-bones, etc. So the idea of a leather lining, for horses doing much road work is good. Nearly all driving horses wear thicker shoes than riding horses because they do more road work and need more protection from bruised soles.

Shoeing is very important to all horses whatever their work. The necessity of a good farrier is ever present with any horse owner. The scarcity of well trained smiths is constantly before us, therefore most of us support the Apprentices Scheme put forward by the Farriers' Association. Nobody is rash enough to cross that valuable craftsman, the good shoeing smith. Now I feel that the old cry that there are not enough horses to shoe can no longer apply.

BREEDING

Before starting to breed, I think a person should carefully consider whether or not he has the necessary acreage and good enough shelter to be able to keep a mare and her offspring. Has the family enough dedication to be willing to spare the time and money on raising a foal to maturity, knowing that it is a time-consuming operation and can be a tedious job in bad weather? A prospective breeder should next ask himself why he wants to reproduce his mare? Has she good enough conformation? Is she free from hereditary diseases? Or is this just an easy way out because the mare has gone lame, or is too old to go on hunting, or the children have grown out of a pony which has been in the family for some years? Unless the mare has enough about her to justify reproducing from her and to give some hope that there will be a future for her offspring, it is far better to pension her off than to bring unwanted foals into the world to swell the ever-increasing horse population of this country.

Here something must be said about grass management of the land on which the breeding stock is to be kept. The British Horse Society produces a very useful cheap little book *Keeping a Pony at Grass* with good illustrations of suitable and unsuitable fencing, water containers, etc. What the breeder has to consider is that grassland has to be looked after. To be kept in good order and as worm-free as possible it needs resting, manuring or fertilising and then harrowing in the early spring to produce good grass for the brood mares and young stock in spring and summer. There must be water tanks or natural running water with not too steep banks. (N.B. Brood mares should not be left to foal out in a field with a stream as so often they go to drink before foaling and have been known to foal too near the water thus drowning the foal.) Plans must be made soon after Christmas as to which fields will be manured, harrowed and laid up for 2 or 3 months. The fields that continue to be used for turning out must get their turn of rest, manuring and harrowing as soon as they can be spared. To manure grass is to feed the grass. Free tests can be taken by the County Agriculture Services to find out what minerals are lacking in your own particular soil, however, your local successful farmer will, I feel sure, give you good advice on the matter.

A contractor can spread the fertilizer for you. Lime is very good for horse-grazed land. But best of all is good old-fashioned muck—well rotted straw manure. Sawdust is not suitable for bedding on a stud farm. Should it be used it must be burnt before being put on the land. Spraying as a control of weeds is used a lot by the farming fraternity but I feel it is only necessary when there is an excess of docks which need killing. Other natural weeds in the hedgerows should be left there, they are the natural extra source of minerals for the horse. Care must be taken to keep all stock away from hedges or wind-carried spray from the next field as the withered weeds are very poisonous. Cattle manure is the best for horse pastures. The grazing of de-horned cattle with horses is excellent since the horses eat behind the cattle and the cattle behind the horses so that there are no tufts left in the fields. Also, from a worm point of view, the cattle worm cannot live in a horse and vice-versa. A vet once said in a lecture, "the horse worm gets into a cow's stomach and says, 'Heavens! I am in the wrong house!', takes off his hat and dies". I have remembered this for years, a good strong point well driven home. The fields should be cross-harrowed at intervals throughout the year i.e. the harrows dragged east and west and then north and south so that all droppings are dispersed. If the acreage is very limited the droppings should be picked up, but in my opinion breeding should not be undertaken on a very confined space. If you have a nice farm friend as I have in Maurice and his wife Anne, we take his mare every year to foal down and cover and he takes our young stock with his own every year for a summer run. This is excellent. A change of herbage and scenery is like a holiday by the seaside for a child.

The ideal fencing is post and rails but this is very expensive. There is a cheaper way of doing wooden fencing by getting rails and posts from the Forestry Commission or any large estate. They do not look so nice but are effective. Barbed wire and young horses just do not mix. Sooner or later tragedy will strike. Gates should have two fastenings on them. It is amazing how horses can fiddle with catches and fastenings and get out. Double fencing is ideal and according to Newmarket the distance between the fences should be 12 feet.

The next question is, what does the owner hope to breed? The decision on this point influences the choice of a suitable sire. Look carefully at your mare and decide where she could be improved so that, when looking at possible stallions, you know what points you want to correct. Should you wish to improve on the head, or make a better shoulder, then bear this factor in mind when looking at the stallion.

Figs 229-230. Mares and foals at grass

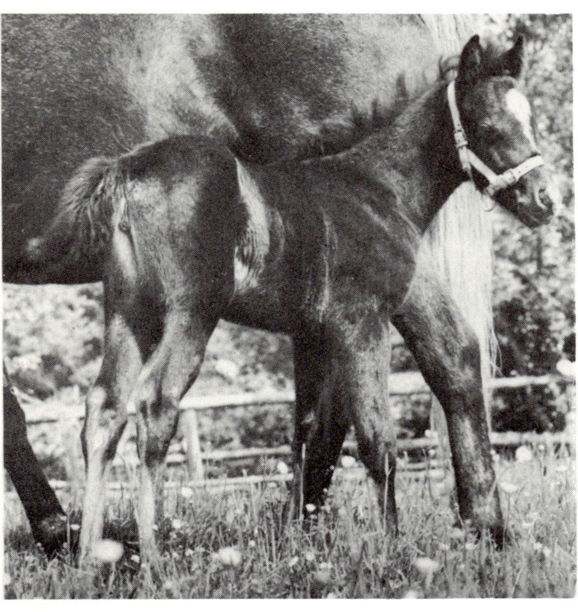

It is always most interesting to look at stallions and to consider what each horse has to offer as a sire. I feel that a stallion must be really masculine in appearance to be dominant in the partnership and to pass on his conformation, action and temperament. This last factor is, I think, the most essential. An easy and kind temperament is all-important because whoever is going to handle and break and eventually own this young horse which is going to be born will want an animal that can be enjoyed. I think this aspect is now considered more than it was 10 years ago. With race horses, on the other hand, temperament is sacrificed for performance. Sometimes these horses, who have great courage, get easily bored when retired from the very active life of racing and get bad tempered through lack of exercise and being confined in the stable.

The visitor should be able to judge whether the horse is really irritable or just inclined to snap when handled. Today stallions are treated much more as ordinary horses than they were in my young days. They are ridden out and exercised daily, which means that they lead a much more normal life and are no longer, as in Will Ogilvie's verse *Bansee* . . . "He stood there chained to the wall, a soul in piteous plight." Always ask to see the stallion out in hand. A horse may look most impressive in his huge box, with lovely rugs on, but may be disappointing in movement and his feet may not be good when looked at standing on level ground. If a horse has broken down when racing and been fired, I always feel that there must be some inherent weakness or a conformational fault. Feet and legs are all-important for a sound horse. But, of course, no horse is entirely perfect. There never was a perfect horse born. However, we are looking for as near perfection as we can get and, in this case, one whose conformation will improve that of our own mare.

Today there are far more stallions kept than there were in days gone by and therefore more imperfect ones. Do not be tempted to use the stallion down the road because it saves trouble. Be careful to see what arrangements there are for your visiting mare; ask yourself whether you feel that your precious mare will receive kind, considerate attention and will be in knowledgeable hands. You must arrange what fees are to be charged and you should try to visit the mare while she is at stud unless you are quite happy about the care she will receive. So often mares are returned desperately thin and worm-ridden and it takes six months to put them on their legs again, besides probably affecting the foal which is being carried. A cheap stud can be a false economy as it costs a lot to get a mare back into condition once she has got really poor.

When arranging about sending a mare to stud, decide what month you wish the foal to be born, if she is a barren or maiden mare; bear in mind that gestation takes about 11 months. So, if you want a March foal, the mare should go to the stud in April. In the case of mares already with foal at foot one is governed by the date of birth of her foal. Most studs like to have the mare on the sixth or seventh day after her foaling. Mares can come on heat, or oestrus as it is called, on the fifth day after foaling but this is unusual. The old accepted period of seven to nine days is usually right.

When sending a mare to stud with a young foal at foot it is important that the stud farm should have boxes available for use in case of wet, cold weather—unless, of course, you are dealing with Mountain and Moorland stock. Even in this case I think there are occasions when they, too, need the shelter of a stable if the foal is very young; if the weather is very bad they cannot find a natural shelter in a field as they can do when running wild on a mountain or moor where rocks and natural terrains can form wind breaks and offer good shelter. A wise, considerate owner will realise that these comforts have to be paid for: no stud can afford to bring in a mare and foal and keep a box clean and bedded down just for the love of horses.

A mare should come into 'use' every three weeks and her period of heat should last for three days. It is usual for her to take the stallion as soon as she will stand for him and be covered again 48 hours later, and to be found off heat again in another 48 hours when she is tried. Normally a stud will keep a mare from 3 to 6 weeks. When a mare is ready to take the stallion she will wink and stale and usually be willing to stand quietly to be served. This is where the stud groom and his assistants have to be patient and understanding since some maiden mares are quickly frightened. If the stallion is noisy in his approach, he may even upset regular breeders. When a mare visits a stud she should be tried daily with a stallion so that she should not be missed. Most studs keep a 'teaser' who should be kind yet encouraging in his approach to mares.* The ideal is for the owner to send his mare as soon as she is in season. Having noted the date of the previous heat he can inform the stud to expect her round a certain date, which will vary considerably according to the mare in question. Some are only 'on' on one day and refuse to stand or take the stallion again. This is the case with one of my own mares who is a regular breeder and only comes into season on one day but has her foal regularly every year after being covered just once. When she went away to a stud, although I told him of this peculiarity the stud groom refused to believe me, and I got a huge bill for keep and vet's fees as they could not get her to take the stallion—this particular stud only tried the mares every other day!

There are other mares who happily take the stallion every 48 hours for a period of over a week. But this is where the stallion owner can get worried and call the vet to see whether all is well with the mare and her reproductive organs. With the spread of breeding diseases and infections I think it is now almost imperative for all stallion owners to insist on a swab being taken from most mares. Besides, when there is some small acidity or infection which would stop the mare conceiving and which could be easily rectified, this simple test can save the mare owner money.

Looking back over the years I wish I had known what I do now about breeding. I might have had many more foals from the mares from whom I first bred. The approach in earlier days was much more one of leaving everything to nature. There was nothing like the veterinary assistance we have today. The fact that the mare had finished breeding was taken as a fact: the inquiring mind did not ask why.

The mare at stud can now enjoy the early grass and be tried and watched carefully in case she should 'return', as it is called, when she comes back into 'use' or 'season' after having been covered. The foal heat is now looked upon as not always being the ideal time for covering. Yet in days gone by it was the one heat everyone felt was the most effective period in which to serve a mare. So the approach to breeding does alter a little as people stop to think. I do feel that in some cases a mare's womb has not settled down enough to accept the fertilised egg and should a mare be torn when foaling I think it gives time for good healing if the foaling heat is missed.

The stitching of the vulva has now become something of a fashion which I do not care for unless it is necessary. Once a mare has been stitched she cannot have a natural birth since the vulva has to be

*The teaser is a stallion who tries the mares daily to see whether they will stand or are ready to serve. Each mare at stud is led so that he can nose her, lick her and generally encourage her. When mares are not ready they show this by kicking at the gate, or laying their ears flat back, bite or even strike at him. It is thought that these teasers get bad tempered and are much to be pitied. As long as a stud uses its teaser for some mares and exercises the stallion well, I think there is no great harm, but they must be well handled and very manageable and checked if they are rough in their approach.

cut to allow the passage of the foal. Should it not be cut it will tear very badly so that in any case it will need re-stitching after the next foal has been born. On the other hand, I think now that many mares should have been stitched in the past but were not unless they made a noise when the intake of air occurred; they were known as 'keyhole' mares and were stitched in the same way as many mares today. But I do not agree with maiden mares being stitched unless they have been torn by a rig or by accident. I feel that the stitching of a filly's vulva is unnecessary interference with nature. In other words, 'How silly can we get?'

The stud will take special care to try the mare around three weeks after she has been covered as it is the natural cycle of her heat period, and again at six weeks in case she breaks. After this, she can be tested in foal before leaving the stud if the owner wishes or she can go home untested and hope that all is well.

There are always two views to be considered over this early testing of in-foal mares. The margin of error is very small. The mare can absorb the foetus or a tiny foetus can be lost and nobody is any the wiser. Some mares will not show that they are in season later in the year even if they are empty. Other mares have been known to show in season and yet be safely in foal. There is always the 99th case in 100 that proves the rule wrong or the accepted theory untrue at times. Mother nature is a lady of whims and fancies.

Once your mare has come home you can ride her quietly but I, personally, never do. Unless there is a good reason why not, I try to turn my mares out into a quiet field away from other horses to enjoy the summer sunshine and to be lazy under the trees. They must have shade, good grass and fresh water and if they look as if they need an extra ration they get it. A gross mare must be kept on shorter grass than the Thoroughbred type. The daily visit to the in-foal mares is very pleasant, as well as necessary.

When winter comes I feel all mares in foal should be kept dry and warm at night; I see no point in turning them out in pouring rain unless they have been in the stable over 48 hours, when I feel a New Zealand rug and some exercise is necessary. All the time the mares have come in at night, they should be getting a short feed night and morning with as much hay as they wish to eat unless they are carrying too much weight, not in their stomachs but over their quarters and in their necks. The size of a pregnant mare's middle varies greatly. Some lose their figures, as do humans, very markedly. Others retain their trimmer figures until late on in pregnancy. I have often wondered if a certain mare could be in foal in December, knowing that she had been successfully tested in foal earlier, but later she has dropped and foaled.

November, i.e. about the 7th month, is the time when mares might slip their foals. Fortunately for me, only once have we experienced this. Luckily the mare was found a very short time after she had aborted. The dead foal was still very warm and the mare needed attention since she was bleeding. You can have your mare implanted to guard against abortion, but I do not feel that this should be necessary unless the mare has a history of losing her foals, in which case I would have her implanted. The vet inserts an implant into the neck to stop the mare's ovaries working, i.e. so that periods of heat will not occur and thus the likelihood of the womb opening will be reduced. The implant should be removed just before foaling is due.

The feeding of a brood mare should be simple and wholesome. Natural exercise at grass and restricted exercise in bad weather keeps the mare well in herself. She should have good oats and bran which are, in my opinion, the staple diet of a horse just as meat and bread is for man. A lot of things can be added and, with artificial manures today locking up so many natural minerals, I think some added proprietary brand of minerals, such as Equivite Bone Meal, should be given, besides having mineral licks in their boxes. Some horses enjoy sugar beet pulp which gives a sugar content. Most horses like flaked maize which can be fed in small quantities dry (a double handful) or scalded added to a night feed which makes a nice warming feed for a chilly evening. Barley, fed boiled or rolled, is another good additive. It is warming to the blood in winter and encourages condition in the mare, but it should not be fed to a mare whose blood is inclined to get over-heated, i.e. if she has spots or heat bumps under the skin. I personally use milk equivalent or calf milk. This is rich in protein but has to be carefully fed and can cause trouble if over fed. A little added to a feed at night and morning works wonders on any animal whose condition is not quite as good as you want it to be.

Plenty of chaff makes horses eat their short feeds slowly and this means that they get the maximum benefit from the food put into their mangers. This is particularly important in the case of brood mares whose stomachs are getting full and for whom good digestion is essential. The gobbling of food can only produce indigestion when unmasticated food passes into the stomach. I like my mares to take a good half hour eating in their mangers before they get turned out in the mornings, then they go out comfortably full to walk and graze at leisure.

Good hay is essential in order to give the mare

bulk in her diet. I think a mixture of seed and meadow hay is the best since the seed hay provides the protein in the bulk while the meadow hay gives palatability and some minerals in the variety of grasses in its makeup. Each grass is known to contain different minerals.

As the time gets nearer to the mare's date of foaling, a careful watch must be kept on her to see that her bowels are kept open and that she is well in herself. She should have an anti-tetanus booster to pass immunity on to her unborn foal. Any signs of approaching foaling must be noted and reported to her owner or the stud groom; they may be false alarms but it is much better if too much is noted instead of the attitude 'It didn't appear to mean anything'. Each mare differs in her approach or symptoms of foaling and any change in her appearance should be watched for, especially her udders and the milk vein.

Foaling

Some people view this natural cycle of reproduction with grave concern. It is true that things can go wrong, and sometimes human assistance is needed, but on the whole nature is able to look after this normal physical process, especially in our native breeds. However, we live in an artificial world these days, and we humans have interfered with nature to suit our own ends, whims and fancies. Over the years we have bred, under artificial conditions, a race of horses in England known as the Thoroughbred. We have played with nature and purposely produced foals in the winter months to enable these horses to race earlier and to have the advantage of these few extra months in the year of their birth before running as two-year-olds. So studs cater for these early births and have to produce warm conditions for the mare to foal in when there may be snow outside the stable. Nature meant mares to foal in late spring when the weather is more clement, though it must be admitted that a mare will foal quite happily on a frosty morning–one will find mother and child quite happy and they will not suffer in any way; at least, this has been my experience over the years. But I think as time has gone by I have become more anxious over my mares, and for that matter more worried over other people's, particularly since I have kept a stallion and run a small stud. So I have tended to alter my attitude toward artificial care during parturition (to give the event its correct name).

In my early days of breeding I used to visit the mare every two hours in the field. After foaling, if I could catch or get near the mare, I would bring the mare and foal into a prepared box. One very wild mare from Ireland I could never handle for a week after foaling, nor could I milk her out on weaning–yet she was quiet to ride.

Now the routine at foaling time at my stables is for someone to sit up doing 2-hourly spells of duty and visiting the mare every half hour. The person on duty calls the next one, and so on, until the one taking over at 6.0 a.m. starts feeding and mucking out. (Those who have been on duty during the night after 10 p.m. can sleep until they wake up.) The person on duty has instructions to come straight for me on the first signs of foaling, i.e. when the mare is getting up and down, sweating or the water bag is showing. Should we have students, either veterinary or N.P.S. Diploma, they are called to see, but usually just the person on duty and I are there to help–if necessary. It always amazes me to find how few students have been present at the birth of a foal. I feel it is important for everyone who is going in for breeding to see a normal birth, so that they can recognise when things are going wrong.

The usual signs of the approach of foaling in a mare are waxing over, i.e. small globules of wax-like material appear on the teats: she may run milk and the muscles by her tail will be relaxed. On the other hand there may be no signs at all and she will foal unexpectedly. A normal birth of a foal will take anything from five to twenty minutes. The mare may foal lying down or standing up. Some mares are quick and peaceful: some are excited: some show signs of distress and panic. Every mare has her own ways and one gets used to one's own mares and their special habits. Some vets and owners like to be very careful hygienically; others prefer the mares to foal outside in more natural conditions. A young mare seems to need less care than an old one, but I do think that we owners become a bit fussy with our favourite mares and worry over-much.

The sequence of events at a birth are as follows. First the water bag appears. As the mare strains it ruptures and there is a gush of water from the bag in which the foal has lain. With this the two front feet should appear, one is often just a little more advanced than the other. Then the nose appears, lying between the legs. As the mare strains the whole head appears. If you are at hand, clear the nostrils of the foal, slap the face and make sure the foal is breathing. A finger in his mouth will take the valued 'foal's muff' out and a cough from the baby will show that all is well. Soon the shoulders are through the pelvic arch and, after that, the whole body and the hind quarters slide through with no difficulty. The mare may stay lying down for a moment or two and then get up, thus rupturing the cord. Alternatively, the foal may struggle and rupture it itself. There should not be any undue bleeding from the cord, just a few drops of blood; should the cord show any signs of bleeding, then tie it with the already prepared sterile tape to stop any unnecessary loss of blood from the young body.

When the mare is standing up, knot up the after-birth, which will be hanging from her, so that she does not tread on it and cause a haemorrhage by tearing it from the sides of her womb. Also the weight of the hanging membranes helps the after-birth to come away naturally. When the mare drops the after-birth, pick it up and examine it to see that all the horns are complete. Should there be an incomplete bag or torn sides of the membranes, then one knows that the mare has not cleansed properly. Should this happen, (or should the mare not be 'cleansed', as it is called) the vet must be called within four hours of foaling. He may be able to help it away manually or he may give the mare an injection to act on her womb to make it drop naturally in a few hours. He will leave instructions as to what to do.

Some mares never cleanse naturally nowadays, but with the aid of modern drugs there is no great need to worry. In the old days the limit was four hours before one worried about the welfare of the mare, and feared the setting in of the dreaded mortification. A cow can retain its cleansing for two days without harm, but not a mare.

This year a mare, which appeared dirty to me, came to stud. I had her examined and it was found that some cleansing, though only a little, had been retained up to the tenth day after the previous foaling. This shows the necessity of laying out a cleansing and examining it carefully before throwing it into a skip to be buried.

Having seen a normal birth, then one is able to recognise any difficulties that may arise. It is a great mistake to interfere too much; on the other hand, a judicial helping hand can be very useful. A slight easing of the shoulders can save the mare a lot of trouble and unnecessary straining. Also, of course, there are times when a foal is malpresented, and in this case veterinary help will really be necessary. Even after the vet has been rung it is very important that everyone remains as calm as possible and waits for his arrival, preparing everything he can possibly need. A table outside the foaling box for his equipment; a towel; a mild disinfectant soap; a bucket, basin and kettle of boiling water at hand. Lights switched on at the yard so that no time is lost through his not being able to see where to park. Someone ready to carry his things from the car. These half-hours seem hours when waiting.

Before I started breeding from my own pony mares I had a small herd of dairy cattle, so at first I assisted at many calvings and then used to calve my own cattle down myself. This has helped me a lot in my approach to foaling, bearing in mind that cattle can take a very much longer time to calve down than a mare will to foal, since a mare gives up trying very quickly. Once the spasms, or contractions of the womb, have ceased, it is much more difficult to deliver a live foal.

But it is amazing how foals can survive. This example will give everyone heart. Two years ago we had a malpresentation. It was a maiden Thoroughbred mare who was not a very calm subject at the best of times, but somehow after the first panic she settled down to being helped. My vet did a wonderful job. We had many helpers and all ended happily. The conditions were that one foot only appeared after the water had broken. I could not find the other foot. The mare was waltzing around and I have the grave disadvantage of being very small myself. So I rang the vet. I tried to keep the one foot back in the mare as much as possible so that the cord would not break, but I was very glad to hand over to the vet when he arrived. At first there were grave doubts and fear that the foal might have to be cut up inside the mare to save her. But at last the other foot was found tucked up behind his head. The two feet appeared after a lot of trouble but still we could not move the foal. We had a rope fixed on each fetlock and a man and a girl on each rope pulling as the mare strained. On the vet's command 'pull', we eased the head through. He slapped the foal's face, breathing into its mouth after clearing the muff and membranes. At last it lay on the straw. My vet turned to us all, saying 'Now we have got it out, don't let it die.' It was amazing that it was still alive after all that pulling and trouble. Elaine and I stayed on after everyone else had gone to bed. We couldn't make it suck. It would go limp on us when we tried to help it to the udder to suck and it collapsed in our arms. So we milked the mare and gave the foal colostrum three or four times. The mare was happy licking its struggling youngster. In the end we were so tired that we both fell asleep in the straw in a corner of the box. I awoke hearing a sucking noise! The foal was sucking on its own. It was getting light at 5 a.m. Another day: another young life on its way into the world, and I made the usual cup of tea . . .

The malpresentation of a foal is best dealt with by a veterinary surgeon. And very clever these gentlemen, and ladies, can be. The great thing is to recognise that all is not well and to know when to call the vet. I think it is far better to call a vet unnecessarily than to leave things and hope all will be well, when delay may be fatal. A mare is unlike a cow; she gives up more quickly and ceases to strain when she finds her efforts ineffective. But these efforts must not be confused with her preparing to foal. She may get up and down several times and be most uneasy before the water bag even shows. It is after the water has broken and nothing appears that one should begin to consider what is going on. Then is the time to feel carefully. Can one feel two feet? And where is the nose? Is the head thrown backwards stopping the feet, legs and shoulder passing through the pelvis?

Firstly, if you yourself have never delivered a calf or other animal leave well alone and rush for the telephone to call for help. Should your vet, who should have been warned previously that you might need him around such and such a date, not be available, nor any of his partners (which is highly unlikely), then my advice would be to call a neighbour who has cows. Roughly, all malpresentations have to be pushed back into more space so that, without the restrictions of the pelvic bones, the foal can be turned into an acceptable position for its passage outwards through the pelvic arch. This is far too much for most laymen and usually one's vet comes within half to three quarters of an hour. Your main concern while waiting for his arrival is to keep the mare as calm as possible and to see that all is ready for the vet's arrival.

There are roughly five main malpresentations and variations of each one which give problems. The first is with a normal position but the front legs are wrong. The second is the variation of a breech presentation. The third presents problems of the position of the head. The fourth is when the foal is upside down. The fifth is when the foal is turned round and is not lying parallel with the mare's backbone.

The easiest of these malpresentations is when one leg is bent back from the knee. The foal must be pushed back so that there is room enough for the leg to be drawn forwards into the natural position for birth. The next easiest is when both legs are bent back at the knees. Again the legs must be straightened out and brought to normal presentation position. The next to be considered is when all four legs are presented. Here sometimes the vet can put cords on the forelegs and push the hind legs backwards, but on small mares this is very difficult. The length of delivery endangers the foal's life. The third malpresentation concerns the head, which may be tilted sideways or downwards or backwards and upwards. In any of these cases, it could not pass through the pelvic arch without the vet bringing it into the correct position.

The three positions of breech presentations are always dangerous since the foal's head is the last part to leave the mare; therefore the chance of suffocation is more likely. The easiest of these is when the hind legs are presented and strong traction may save the foal. But when the buttocks are presented, the hind legs must be drawn forwards and great care taken to see that the cord is not broken, breaking the oxygen supply from the mare to the foal. The last of the 'breech presentations' is when the foal lies with all four legs close together. This again takes more time and sometimes necessitates the cutting of limbs from the foal before it can be delivered. But now we often hear of a Caesarian section being successfully carried out.

The upside down presentation can be turned into a position for a normal birth if all the factors are in the vet's favour, namely:

(a) the mare is a large one and stands still;
(b) the foal is small;
(c) the cord does not get in the way and get broken.

Likewise the breech presentation upside down can be made into a normal breech presentation, as can the dorsal presentation, i.e. the middle of the back, by opening the womb. This can be put right or may need dissecting before the foal can be removed from the mare. Another, and probably the worst malpresentation, is when the foal is curled round, presenting four legs and the head with the nose and hocks lying together. This position needs a great deal of rectifying. In rare cases a live foal can be delivered, but it is unlikely.

Please do not dwell too much on these abnormalities in parturition. They are rare. Modern science and our vets' skill have advanced greatly, making owners' difficulties less each year. Foaling is a natural process of life. The female gives birth. It was meant to do so. It is a wonderful fact of nature. In most people's opinion, the birth of a foal is more likely to go well unaided by man, but someone should, in my opinion, be there to help in case things go wrong and the mare need help. Please note the word *skilled.* I have never known a vet complain at being called out in the middle of the night, even if the foal is standing when he arrives. A polite apology plus an explanation and self-condemnation always go down—plus a whisky in the tea! I have had many a good conversation over the kitchen table at 2.0 a.m. And to those kind gentlemen who have answered my calls for help and been so kind to me, I would like to say 'thank you very much'.

VETERINARY

There are many very good and some excellent books written on the veterinary aspects of dealing with horses. Two which come to my mind are Hayes' *Veterinary Notes for Horse Owners,* a very full, descriptive book and a little above the ordinary owner, but most useful to keep on the shelf for reference; the other is a recent publication called *The T.V. Vet Horse Book,* dealing with many common ailments, beautifully illustrated, very well written and it cannot fail to be of great use to any horse owner. Here I am only going to touch on two or three of the most common complaints which most horse owners will come across at some time or other.

Laminitis is very common with children's ponies and is caused by a pony becoming too fat on the good summer grass. It is inflammation of the sensitive laminae of the feet. The treatment is to reduce the ponies' intake of food containing high protein, i.e. literal starvation in most cases. The horse should be given bran mashes with epsom salts added (to take away the unpleasant taste of epsom salts, a little black treacle may be added, diluted in hot water to make the bran mash really sloppy). Should a stream be near at hand, stand the pony in it to cool the over-heated feet. If taken in time the pony will probably be able to return to work within a week, but when really bad the pony may have a dropped sole and, in severe cases, it is the finish of its useful life.

Thrush can cause lameness. It is usually caused by the feet not being picked out regularly and an evil-smelling black liquid forms, causing painful places between the frog and sole of the hoof. This should be washed out with disinfectant and Stockholm tar painted into the foot. Often the horse will not be lame if this is discovered in time, hence the desirability of picking the feet out daily.

Another common complaint is of a horse whose stomach becomes a little upset and his droppings become too loose. This can be corrected by reducing oats and any green hay and feeding him on dry bran (which will bind up the stomach) and brown hay. If he fails to show any improvement, a vet should be called.

Constipation is another common complaint in a stabled horse and this can lead to a stoppage if not dealt with. Give bran mashes with hot water, to which oats may be added if the horse is not keen on bran alone. His hay should be a green rather than the brown variety and can be damped to give more moisture to his stomach. Linseed tea may be added to bran mashes, but if there is little improvement in his droppings, rub two tablespoonsful of linseed oil (not boiled linseed oil as used for paints but veterinary linseed oil which can be obtained from Boots) into the bran with which you are going to make his bran mash, again adding treacle to take the taste away so that he will eat it readily.

Regular *worming* of all young stock is most important. Your vet's opinion should be obtained about worming in general. The brood mare should be wormed according to your vet's advice as everyone has a different theory. Mares should be given a small amount of worm powder for the first 20 days in every month to keep the egg production of the female worm at a minimum for the sake of the pasture. But this is not a worm eradicator for the mare in any way.

Books recommended for further reading:-

Breeding

Mares, Foals & Foaling by Friedrich Andrist, published by J.A. Allen, London.

Veterinary

Veterinary Notes for Horse Owners by Captain Horace Hayes, revised edition published by Stanley Paul Ltd, London.

The TV Vet Horse Book published by Farming Press Ltd, Ipswich.

Riding

Equitation by Harry Wynmalen, published by Country Life, London.

Riding Logic by W. Meuseler published by Methuen, London.

From Paddock to Saddle by E. Hartley Edwards, published by T. Nelson & Sons Ltd, London.

Equitation published by The British Horse Society, Warwickshire.

Photographic Acknowledgements:-

Most of the photographs were especially taken for this book by Ian Gibson-Smith and Paul Turner of Design Practitioners Ltd, Sevenoaks. Other photographs were obtained from:–
Photonews–*Figures 161, 167, 169, 170, 171, 172, 174, 175, 176, 180, 181, 182.* R. Clapperton–*Figure 177.* Findlay Davidson–*Figure 84.* John Topham–*Page 74 & 85.* Sport & General–*Figure 165.*

Index